THE THREATENED GENE

The Threatened Gene

FOOD, POLITICS, AND
THE LOSS OF GENETIC DIVERSITY

CARY FOWLER AND
PAT MOONEY

THE LUTTERWORTH PRESS
CAMBRIDGE

The Lutterworth Press
P.O. Box 60
Cambridge CB1 2NT

British Library Cataloguing in Publication Data
Fowler, Cary
The threatened gene
1. Agriculture. Use of genetic engineering
I. Title II. Mooney, Pat
631.523

ISBN 0-7188-2830-5

First published in the United States by the University of Arizona Press

Printed in Great Britain by
Billing & Sons Ltd, Worcester

To the next generation
Robin, Kate, Sarah,
Jeff, Nick, Morgan,
and Joel

CONTENTS

INTRODUCTION

While many may ponder the consequences of global warming, perhaps the biggest single environmental catastrophe in human history is unfolding in the garden. While all are rightly concerned about the possibility of nuclear war, an equally devastating time bomb is ticking away in the fields of farmers all over the world. Loss of genetic diversity in agriculture—silent, rapid, inexorable—is leading us to a rendezvous with extinction—to the doorstep of hunger on a scale we refuse to imagine.

To simplify the environment as we have done with agriculture is to destroy the complex interrelationships that hold the natural world together. Reducing the diversity of life, we narrow our options for the future and render our own survival more precarious. It is life at the end of the limb. That is the subject of this book.

Agronomists in the Philippines warned of what became known as southern corn leaf blight in 1961.[1] The disease was reported in Mexico not long after. In the summer of 1968, the first faint hint that the blight was in the United States came from seed growers in the Midwest. The danger was ignored. By the spring of 1970, the disease had taken hold in the Florida corn crop. But it was not until corn prices leapt thirty cents a bushel on the Chicago Board of Trade that the world took notice; by then it was August—and too late.

By the close of the year, Americans had lost fifteen percent of their most important crop—more than a billion bushels. Some southern states lost half their harvest and many of their farmers. While consumers suffered in the grocery stores, producers were out a billion dollars in lost yield. And the disaster was not solely domestic. U.S. seed exports may have spread the blight to Africa, Latin America and Asia.[2]

The real culprit was not the disease but crop uniformity. As a U.S. National Academy of Sciences publication later recalled, "from Maine to Miami; from Mobile to Moline" virtually all commercial corn varieties were genetically identical in at least one respect.[3] When one genetic component of most varieties became susceptible to the new blight, the whole American crop was vulnerable.

In the autumn of 1971, farmers in the Ukraine settled into the Russian winter comfortable in the knowledge that their fields were seeded with Besostaja, the highest-yielding wheat the region had ever seen. As January temperatures slid lower and the much needed snow cover failed to materialize, fears of winter kill spread. When the spring rains also failed, farmers and politicians alike knew the July harvest would be poor.

In April, 1972, about the time the National Academy was wrapping up a landmark study on southern corn blight and the genetic vulnerability of other crops, American Agriculture Secretary Earl Butz was in the Ukraine touring the wheat fields. Although members of the delegation were aware of the shortage of rain, their attention was on the larger spring crop, not the winter growth. Based on what they saw, U.S. officials concluded that the Soviets would not be big buyers of grain that year.

Georgina Vitonova thought otherwise. Back in January, the Ottawa-based economist had penned a report to the Canadian Wheat Board arguing that the Russians were in for a disaster. She had been monitoring the Soviet press—especially the weather reports—and she was aware that the Besostaja, highly responsive to inputs, occupied forty million hectares from Kuban to the Ukraine. She reasoned that it would never survive such a harsh winter. Vitonova estimated that between thirty and forty percent of the winter wheat crop[4]—at least twenty million tons—was lost.[5]

Faced with losing their immense herds of cows and hogs, Russian politicians opted to buy their way out of the crop failure. By February, the Canadian Wheat Board had cut a secret deal on grain imports.[6] In July Russian traders at the Manhattan Hilton were well on their way to having purchased twenty-seven million tons of grain. The world has never been the same since.[7] Grain and bread prices soared. Between Butz's April Ukrainian sojourn and October the same year, the Rotterdam price for a metric ton of wheat jumped from under sixty-five dollars to ninety.[8] North American farmers thought they had died and gone to heaven. Before they came to, a generation of farmers took up Butz's challenge to "get big or get out" and raced enthusiastically into debt buying more land, bigger combines and all the fertilizers, center-pivot irrigation pumps, and pesticides their land could absorb.

If, at first, this was good news for farmers, it was bad news for the world's hungry, who were unable to buy their way out of the same drought that had devastated the Russian harvest. They could not compete with Soviet cows for the high-priced wheat. Between 1972 and 1973, Third World grain imports rose twenty-five percent, but their cost doubled to six billion dollars.[9] The triple blows of Russian crop failure, Sahelian drought and international oil crisis—(aided and abetted by U.S. sabre-rattling about the "food weapon," as Butz termed American agricultural abundance in a world of food shortages)—propelled world leaders into food politics in earnest.

The full and final effects of the Besostaja wheat's collapse—the hunger in the Third World and the engineered boom and bust in the industrialized countries—are still reverberating in world agriculture today. Why did the Russian wheat fail? Just as with the American corn crop two years earlier, the underlying problem was genetic uniformity. Forty million hectares of Soviet soil had been sown to a single variety. High-yielding in the mild winters of Kuban, it was incapable of surviving the sometimes harsh winters of the Ukraine.[10] American corns, by contrast, were vulnerable to a different type of stress—the blight disease.

The epidemics of the early 1970s served to underline a simple but humbling point: although the "North" (meaning most northern industrialized countries) is grain-rich, it is gene-poor. Wherever the Garden of Eden might have been, the Horn of Plenty is definitely in the tropical and subtropical southern latitudes. Maximum genetic diversity is found in the tropical latitudes. While the vegetative assets of the temperate zones were literally frozen during the ice ages, botanical diversity flourished in the warmer tropics. As people later moved from the tropics, they took their seeds with them. Those who first crossed the oceans packed a lunch. The genetic "homes" of the thirty crop plants that, in aggregate, give humanity 95 percent of its nutritional requirements are all to be found in Asia, Africa, and Latin America. Were one to list the top five crop species for every country, only 130 crop species would be named, virtually all originating in the Third World.

Despite popular misconceptions, corn did not originate in the United States, but in Mexico. The solution to the southern corn leaf blight—other than through a wider breeding program in general—was finally found in Mayorbala maize from Africa (though it too must have originated in Central America). Similarly, the Russian search for winter hardy wheats took them from the Fertile Crescent to the Himalayas.

Seeds are unique in that the means of production—seed—is often also the end product for consumption. The rapid replacement of old "farmer" varieties with new "scientist" varieties can hasten the demise of the old genes. Modern plant breeding began in the twentieth century. By the end of World War II, almost all of the enormous number of wheats grown in Greece had been replaced by a handful of new varieties. As the mid-1970s were reached, three-quarters of Europe's traditional vegetable seed stood on the verge of extinction. By that time scientists were beginning to scrape the bottom of the barrel—in this case the gene pool—in search of genetic resistance to an ever growing list of virulent diseases and menacing pests attacking the world's most important crops. Although modern breeding had led to a "green revolution" in the North and a massive boom in yield, it had also eroded the genetic base for future breeding. We had, as Garrison Wilkes of the University of Massachusetts pointed out, built our roof with stones from the foundation.

In this book we frequently use the geopolitical terms "North" and "South" and "Third World." As a rule, when we refer to the North we are talking about the industrialized countries of Europe and North America. We would also include Australia, the USSR, and the nations of Eastern Europe in this category, though they are often of only minor importance in the subjects under discussion. We obviously do not mean to use the term North in any strict way; using the term loosely is preferable to laboriously listing each country every time we need to employ the term. Similarly, the South includes the countries of Asia, Africa, and Latin America, which others call developing or underdeveloped. China, by choice, prefers to be considered in this category, and we use the terms with this in mind.

The North's genetic dependence on the South (the tropical areas of the Third World) is accelerating for many crops. In 1970, direct input of Third World germplasm contributed about a quarter of the North American spring wheat crop. By 1983, the continent looked South for fully half of its entire wheat breeding stock—including the much larger winter wheats.[11] The Paris-based OECD (Organization for Economic Cooperation and Development) has estimated the value of the South's wheat genes to U.S. agriculture at $500 million a year. By comparison, the U.S. corn crop was relatively independent until 1970 when the use of "exotic" germplasm was probably less than one percent. Blight has now pushed tropical gene imports much higher. U.S. seed industry figures indicate that one potentially useful gene from the Third World may contribute a billion dollars to the agricultural economy. Third World germplasm now contributes over two billion dollars a year to the farmgate value of U.S. wheat, rice and maize.[12]

It was not by accident then that in the spring of 1972—as the world's two superpowers contemplated the fragility of their food supply—an international meeting was convened on the outskirts of Washington to create a United Nations–affiliated network for global collection of Third World crop germplasm. And, coincidentally, on America's other shore, scientists in the San Francisco Bay area were conducting the first gene-shifting experiments[13] that would turn agricultural genes into the raw materials of one of the most powerful industries the world has ever known. It is the confluence of these two streams that form the politics of the world's gene pool today, giving new meaning to the term "genetic resources."

In their collection efforts, northern governments are encouraged by seed companies. Patent monopolies and global opportunities have turned the old seed houses into transnational genetics supply corporations. Genetic engineering is showing the way to still greater profits. The building blocks for the new biosciences are genes. The more genes, the more opportunities to develop new varieties, new crops and new controls over the food system. Our research shows that in less than two decades, close to one thousand traditional seed companies have been absorbed into the fold of a new breed of international biochemical enterprises.

By the late 1980s, the struggle for control of breeding material—seeds, and the genes inside them—has become intensely economic and political. Both nations and companies now vie for access to and benefits from the world's germplasm.

Behind the politics and profits is a history which begins with the hunters and gatherers of twelve thousand years ago and runs to the gene-splicers of today. It has been a long time since women first learned to control the shattering of seeds. Early grains "shattered." The seeds did not cling to the plant but were easily dispersed. But in those stands of wild grains some plants were different. As a result of minor genetic differences, some held on to their seeds. Normally this was dysfunctional for the plant. But to the early "farmer" it was a boon, enabling her to collect seed more easily.

Harvesting and subsequently sowing seeds that remained on the stalk encouraged the non-shattering trait and meant that less seed shattered and fell to the ground before harvest. Harvesting and sowing non-shattering seeds led to the domestication of our food crops and the tremendous diversity found within them.

But with the discovery of modern genetics a few decades ago, the food system began to experience rapid change. With the arrival of the green revolution, the world's food supply has been faced with a new wave of genetic erosion. And with the coming of plant and gene patenting and the opportunity of monopolization, international companies have attempted to corner the market for the vanishing genes. The result may be the shattering of agriculture itself.

What is at stake is the integrity, future and control of the first link in the food chain. How these issues are decided will determine to whom we pray for our daily bread.

A WORD ABOUT VARIETIES

Whether popular or scientific, most published work about genetic diversity shares the problem of ambiguity in the term "variety." To most people, the word "variety" means variation or diversity. To a plant breeder it means something particular; to a wild plant taxonomist, something else again. Technically speaking, a cultivated plant variety is a distinct, named, rather uniform, modern creation also referred to as a cultivar. Scientists do not often speak of "peasant varieties," or the "varieties" of five thousand years ago; these they call "landraces." Landraces are usually more variable, less distinct and less uniform. The same landrace may go by different names in different countries or among various communities. A landrace may express tremendous variation in a single field in height, days to maturity, even pest resistance.

For a botanist or taxonomist studying wild plants, a variety means a geographic race or regionally dominant variant of a species, similar to an animal subspecies.

In this book, since it is for lay readers, we gravitate toward the common usage, as opposed to these scientific uses of "variety." Accordingly we ask scientists to give a little ground. We hope the intended meaning of the word "variety" will be understood by the context in which it is used. When we speak of primitive varieties, we obviously do not mean something bred by a modern plant breeder, nor do we mean to imply that the primitive varieties (landraces) share the characteristics of uniformity and distinctiveness and stability with "modern" varieties.

As this book went to Press, we began to realize that the term "primitive varieties" is an inappropriate one. It tends to denegrate the real and valuable achievements of peasant farmers and downplays the importance of varieties themselves. Given an opportunity to rewrite this book, we would use a new term such as folk seeds or folk varieties to reflect the ongoing contribution of grassroots communities to the creation of valuable and useful genetic diversity.

PART ONE
Legacy of Diversity

CHAPTER ONE

Origins of Agriculture

History celebrates the battlefields whereon we meet our death, but scorns to speak of the plowed fields whereby we thrive; it knows the names of the King's bastards, but cannot tell us the origin of wheat. That is the way of human folly.

—Jean Henri Fabre

For all our technological wizardry, we human beings still owe our existence to a few inches of topsoil, an occasional thunderstorm, and a handful of crops.

Few of us ever pause to wonder how agriculture came to be or why it even exists at all, why people abandoned a life of hunting wild game and gathering wild plants to till the land, sow seeds, and harvest crops. The ancients were less cavalier about their food. They knew then as we should know now that one does not take for granted that which provides and sustains life.

Our ancestors created rich mythologies to explain the beginnings of agriculture. Each culture had its own unique story to tell. The Turco-Tatar people of the Middle East, for example, believed the teacher of agriculture and the inventor of fire to be a wise and crafty porcupine.[1] The Babylonians explained that the god Oannes came from the seas and had the body of a fish, but the feet of a human. They believed that this deity taught people science, art, and architecture, and "introduced agriculture and all which would soften their manners and humanize their lives."[2]

In Chinese mythology, Shen-nung, an ancient ruler with the head of an ox and the body of a man, taught people agriculture and the use of fire. The goddesses Ceres in Rome, Demeter in Greece, and Isis in Egypt were credited with agriculture in the Mediterranean region, while according to the Aztecs, Quetzalcoatl, in the form of a black ant, stole maize in order to give it to people.[3]

Among the Hebrews, however, agriculture was seen as a curse. The Book of Genesis tells the story of the sin of eating the fruit of the Tree of the Knowledge of Good and Evil. Angered, God says to Adam:

> 'Cursed is the ground because of you; in toil you shall eat of it all the days of your life; Thorns and Thistles it shall bring forth to you; and you shall eat the plants of the field. In the sweat of your face you shall eat bread till you return to the ground for out of it you were taken . . .' therefore the Lord God sent him forth from the Garden of Eden to till the ground from which he was taken.—Genesis 3:17–23

And with that decree, the idyllic lives of Adam and Eve as gatherers in the Garden of Eden abruptly ended.

While such myths of agriculture's origins flourished among farming peoples, earlier hunting and gathering societies had created their own tales giving their lives and livelihood a divine origin. Harkening back to the beginnings of hunting and gathering, a contemporary Aborigine woman explained: "Ngalgulerg gave us women the digging stick and the basket we hang from our foreheads, and Gulubar Kangaroo gave men the spear-thrower. But that Snake that we call Gagag—taught us how to dig for food and how to eat it, good foods and bitter foods."[4]

When Europeans first made contact with this woman's ancestors in Australia, they encountered a continent of three hundred thousand people who practiced no agriculture and had no domesticated plants. They hunted, fished, and gathered wild tubers, berries, fruits, and greens.[5] And they "managed" wild plant populations. Today, such hunting and gathering societies are nearly extinct. They have come to the end of a very long road.

PRELUDE: HUNTING AND GATHERING

Years ago, as college students, we sat in a large lecture hall waiting for the first lecture of the "Introductory Anthropology" course to begin. The elderly, bearded professor walked in, silently picked up a piece of chalk and began to draw a single line all the way across the long blackboard. When he had finished, he stepped back and drew a vertical line a few inches from the edge. "The long line represents the history of mankind on this earth," he said. "The tiny segment at the end of the board denotes human history since agriculture was first practiced."

It was a perspective to which we were unaccustomed. We were to learn that in the grand sweep of human history, only six percent of the people who have lived and worked, laughed, made love, and died on this earth—only six percent—have lived by farming. Just four percent have lived in industrial societies. The dominant and most enduring form of human survival has been hunting and gathering. About ninety percent of the estimated eighty billion people who have ever lived spent their days on earth as hunters and gatherers.[6]

How would we describe life before agriculture, the life nine out of ten of our ancestors led? Was it one of brutal savagery, constant hunger, and misery? Several years ago we offered an answer to that question when we began a magazine article with the statement, "Human civilization began with the sowing of seeds." It was a powerful opener, we thought. But it was incorrect.

Before the first seed had ever been purposefully sown, hunting and gathering peoples had developed religion, customs and rituals, social organization, art, medicine, language, had lived in huts and hamlets, and had made masks, paints, weapons, traps, lamps of a sort, and fishing equipment—boats, nets, hooks, and cordage.[7] Who could say they lacked civilization?

A relatively recent example may show that appearances—and our own preconceptions about primitive hunting and gathering tribes—can be misleading. Charles Darwin encountered in Tierra del Fuego tribes whose speech he thought was barely human. Moreover, in terms of "material culture" they had almost nothing. They wore animal skins for clothing and they knew how to use fire. Not much more could be said about these miserable creatures, it seemed. But in the 1860s when an English clergyman went to live with one of these tribes, the Yahgans, he recorded a vocabulary of 30,000 words.[8] By contrast, fewer than 850 words make up ninety percent of our everyday vocabulary.[9]

Professor Jack Harlan of the University of Illinois once set out to experience food gathering first hand. He literally waded into a field of wild wheat in Turkey and found that with "no prior training" he could harvest one kilogram of clean grain per hour. A family could collect a year's supply in three weeks "without even working very hard."[10] In Mexico, Dr. Harlan went out harvesting wild corn, again with impressive results. He calculated that a typical gatherer could collect an eleven-day supply in just three and a half hours.

Wild foods were not as scarce in the days of hunter-gatherers as we might imagine. When the state of Israel was established, the new country set about controlling grazing. Soon wild wheat reappeared, growing in "stands as dense as cultivated wheat fields" over broad "non-arable" areas.[11]

In ancient China, a second century B.C. observer claimed that no one starved or feared famine, nor was there even need for putting up large stores of food, for wild foods and shellfish were plentiful.[12] Hunters and gatherers did not have to wait in line at the supermarket. Food was usually abundant.

One hunting and gathering tribe that still lives by traditional means and has been well studied is the !Kung of the harsh, arid interior of southern Africa. (These people have usually been called the !Kung Bushmen; "Bushmen" is now considered a derogatory term given the !Kung by foreigners.) Their typical work day lasts about six hours. The mongongo nut, both plentiful and nutritious, is their most important food source. Despite the seemingly uninviting environment, the !Kung spend only twelve to nineteen hours a week getting food—an activity that includes neither children nor the aged.[13] A working class wage earner in the U.S. spends this much time every week earning enough to buy food.

The diet of hunting and gathering peoples was extremely diverse. In 1936, Elias Yanovsky of the U.S. Department of Agriculture listed 1,112 species of food plants used by North American Indians.[14] But Richard Felger and Gary Nabhan, experts on Native American foods, estimate that Indians even in the driest part of the U.S. Southwest were using 375 plants.[15] Their estimates are that 3,000 to 5,000 wild plants were probably used as food in North America.[16] This diversity very likely gave people a more balanced diet than many enjoy today. More importantly, people who lived mainly by hunting and gathering were obviously not dependent on a single food source, and the many foods they ate came from stable (though not always abundantly productive) ecological systems. This meant that they enjoyed a more dependable food supply than farming can normally produce.[17]

Starvation among hunters and gatherers was rare. In virtually every case with which we are familiar, starvation came about only as the result of interference by other people. When crops fail and when social and economic systems do not protect the poor, people in farming societies starve. "But famine," according to Professor Harlan, "is not recorded among gatherers except when there has been a drastic disturbance by outside agents."[18]

PROFESSIONAL BOTANISTS

The need to care for children helped create division of labor among hunters and gatherers. Men hunted, women gathered. Of the two pursuits, gathering was clearly more important. While the capture of a single large animal might have provided a clan of forty people with meat for two weeks,[19] it was gathering that gave our ancestors a dependable diet—probably about seventy percent of their caloric requirements in the arid tropics. Though it has been generally assumed that hunting provided more food than gathering in the high northern latitudes (above 40 degrees), an American anthropologist studying tribes along the western Canada/U.S. border (45–48 degrees N.) found that even this far north, with plentiful game and declining plant resources, women provided seventy percent of the diet from gathering.[20]

Life was not as simple as picking apples from the apple tree, however. People first had to learn by experimentation whether fruit, roots, seeds, or leaves were edible. When it was none of the above, we must assume that sometimes both experiment and experimenter were terminated. No government agencies existed to warn of the dangers of drinking hemlock.

Once beyond identifying edible portions of plants, people would have needed to invent roasting and leaching in order to eat many foods. Grains would have needed to be ground with grinding stones, which may first have been used for crushing ochre for pigments.[21]

In Asia and Africa, wild yams were not used initially as food, for they were highly toxic. Instead, they were used as arrow poison or they were dipped into streams in order to stun fish and visiting birds and monkeys.

Long before modern botany came into being, primitive people identified all five natural sources of caffeine (tea, coffee, cola plant, cacao, yerba mate and its relatives) and knew that they could be used to alleviate fatigue.[22] Such intimate knowledge of plants by our ancestors enabled hunters and gatherers to survive and flourish from the Arctic Circle to the deserts of southern Africa.

Stripping away our misconceptions reveals that hunters and gatherers were in many respects like the farmers to whom they gave rise. They had rich cultures. They harvested wild seeds and knew what to do with them. They used plants for poisons and medicines. They prepared areas by burning off vegetation to favor certain plants.[23] They protected other plants and even prayed for rain like many of our farming neighbors.

Their knowledge of plants and their intimate relationship with plants—often reinforced by ritual and myth—led them to the threshold of agriculture. But that last step from gathering to the conscious selection and sowing of seeds, and cultivation of plants, did not occur for more than two million years.

WHY AGRICULTURE?

Why, twelve thousand years ago (with a worldwide population of about ten million as opposed to five billion today), did people begin to practice agriculture? In lands of plenty, what would have prompted the transition from an easy life of gathering to the difficult and precarious occupation of tilling the soil? Why agriculture?

While studying the !Kung, Richard Lee put the question to one of these surviving gatherers. The reply was just as simple as the question: "Why should we plant when there are so many mongongo nuts in the world?"[24] Why indeed!

When all is said and done, agriculture provides only a few advantages over hunting and gathering. It allows more food to be grown in less space and thus allows for a greater concentration of population. It favors a more settled life. And it permits human settlement in some areas that would not support many gatherers. But we come to these conclusions from the perspective of the twentieth century. These "advantages" may have meant nothing to men and women living twelve thousand years ago. What then would have driven *those* people to practice agriculture?

We have addressed the prejudice that until recently enabled us to see hunters and gatherers only as crude and barbaric. In the same mode as that inaccurate stereotype, agriculture is seen as such a radical departure from the experience of the typical hunter-gatherer that there could be only one explanation for it: it was discovered, invented by a particularly clever savage. Think for a moment, we are asked, of the extraordinary mind it would have taken twelve thousand years ago to comprehend the relationship between seed and plant. Imagine the foresight of that first sowing of seeds. And imagine how agriculture must have taken hold as the first farmer spread the idea. It has been argued that just such a dramatic discovery would have been needed.

Today it is regarded by most experts as unlikely that agriculture began as a discovery or invention in any one place and spread from there. Studies reveal that indigenous crops were often sown and harvested before crops had been brought in from other areas.[25] Agriculture does not have one birthplace, but many. Some people learned it through teachers from foreign lands; others taught themselves.

In many areas the beginnings of agriculture were perhaps so gradual as to be almost imperceptible as the sophisticated practices of hunting and gathering people were blending into what we consider agriculture. Selective harvesting, the burning of cover vegetation, and protection of favored or valuable plants were evolving into something very akin to cultivation. It did not require genius to notice that pieces of yam or manioc chipped off from the main tuber during harvest would sprout and produce additional plants. The step from harvesting to encouraging or "cultivating" these root crops was a small one.[26]

But for some reason, after two million years of human history, this and similar, simple experiments began to be performed over and over again by thousands of people all around the globe at approximately the same time. Why?

Many scientists believe the answer lies in population growth. Hunting and gathering tribes faced with increasing populations turned to the plants they knew so well and began to cultivate them. The theory sounds logical; it even sounds familiar. Is it not commonplace to hear contemporary farmers exhorted to produce more food in order to feed the steadily increasing population of the world?

But the theory that population growth led to agriculture has its problems. To begin with, agriculture is not a very quick way of increasing food supplies. Anyone wanting more food would have had time to consider quite a few other options for a meal while waiting for crops to sprout and mature. In Tehuacan Valley of Mexico, where several ancient agricultural sites have been unearthed, the local diet contained only up to six percent cultivated plants some six to eight thousand years ago. Such a small amount would not have beaten back starvation. And it is doubtful that population pressures would have prompted the move to agriculture for such a modest harvest.[27] To plant a whole field with seed—something untried and of uncertain outcome—would surely seem quite a gamble if the situation were acute. Better to eat the seeds if you are hungry than to plant them if you are the world's first farmer.

Population growth, however, probably did play a role in causing hunting and gathering bands to divide. New bands might have been forced to migrate into less hospitable regions. Such splinter groups would have taken with them their intimate knowledge of plants. But they would have found their new homes less well endowed with the dense stands of wild grain to which they were accustomed. It is not difficult to imagine them encouraging any sparse growth of familiar grain they found. Indeed, some of the oldest samples of domesticated grain found to date have come from these "marginal" habitats, at the edges of the common natural range of the plant.[28]

In Africa, as the ancient lakes of the Sahara dried up, people were forced to move southward, adding to the population already there. This might have put enough strain on the ecosystem to prompt people to manipulate and encourage plants further.[29] Like the prophets of old, agriculture may well have come from the fringes, the margins of human society twelve thousand years ago.

We consider it unlikely that agriculture commonly began purely in order to produce more food. Planting a few seeds would not have been regarded as the solution to a serious food shortage by someone who had never farmed before. Instead, the first plants intensively cultivated often would have been those that were highly valued but in short supply, or perhaps rare. They would have been plants which satisfied a particular need. Medicinal plants, plants used for dyes, pigments, and utensils, plants used in rituals or for magic, and plants used as poison to catch and kill fish and game would be obvious candidates. Their value would have been great, although the quantity needed perhaps quite small. Cultivation would have produced results worth the time and effort. Amaranths, for instance, provided vivid red pigments used in ceremonies from the Andes to the Pueblo region of the southwestern United States. This use of amaranths was more widespread in ancient times than its use as a grain, for which it is promoted even today.[30]

At Çatal Hüyük, an ancient site of both hunting and gathering and primitive agriculture in Asia Minor, there is evidence that a cattle cult was flourishing nine thousand years ago.[31] Bulls' heads modeled in plaster adorn the walls of a shrine.[32] Powerful religious motivations—rather than a need or fancy for T-bones—may have prompted people to domesticate the fierce, six-foot tall aurochs, ancestor of today's cattle. Similar influences, both practical and superstitious, probably were affecting people's relationship with plants as well, contributing to domestication.

In some areas a decrease in game through climate change or overexploitation would have led people to become more dependent on plants and more sedentary. With men at home more, birth rates would have climbed. In combination, sedentary life, population growth, and a decline in game might have led to increased agricultural experimentation. Food crops might initially have been cultivated haphazardly to supplement the diet. Other foods may have been produced as a byproduct of unrelated activity. The traditional Brazilian method of making manioc meal requires removing the powerful toxins in the plant. One step in this process involves soaking it in a stream—precisely what the ancients did with it to stun and capture fish. Manioc's use as a food could have been discovered following its service as a poison.[33]

Sedentary life also allowed our ancestors to begin to produce and accumulate greater numbers of possessions. Many of the tools used to hunt and to gather wild foods were useful in the early stages of agriculture; harvesting seeds encouraged the carving of additional grinding stones; a good harvest produced the need for drying racks and storage facilities.[34] After sweating over these projects, people likely became less and less inclined to abandon them and return to a wandering life as a hunter-gatherer.

As with all other notions of how and why agriculture began, there are exceptions. Sedentary life did not always lead to agriculture, nor did agriculture immediately lead people to settle down. (In some areas, sedentary life enhanced people's skills at fishing, just as it does today!) Long after people began purposely growing plants in Mesoamerica, they retained their nomadic lives.[35] No one theory about agricultural origins explains all the facts.

When Stone Age men and women settled into an area as fishermen, gatherers and quasi-agriculturalists, they began to alter the environment in substantial ways. They cleared land. They trampled and uprooted existing vegetation. They established dump heaps where "kitchen" refuse and human waste were deposited. In short, they created opportunities for those plants naturally adapted to disturbed areas: they provided a home for weeds.

Views differ as to what a weed is. It may be an "unwanted plant," one "growing where it isn't supposed to"—but one person's weed is another's flower. Consider, for instance, that crabgrass, the scourge of America's suburbs, was cultivated as a cereal grain in central Europe until the 1800s.[36] From a scientific standpoint, weeds like crabgrass are plants adapted to disturbed habitats such as would be found around settlements or where the soil had been tilled. They are "pioneers of secondary succession." Viewed from our ancestors' perspective, they were "camp followers."

Ten thousand years ago, the world was opening up for weeds. They encountered the disturbed habitats they liked and often found these habitats fertilized as well. The late Dr. Edgar Anderson, botanist and geneticist, theorized that women noticed weeds growing around the hut and encouraged those that were particularly productive. To be sure, many of our important crops have ancestors in common with weeds. Other crops have weedy cousins. But did agriculture begin in dump heaps? Like the other theories mentioned, this one has a germ of truth, but it also has exceptions, crops in which development would not have been fostered under such conditions. Rather than being the complete answer, the "dump heap theory" gives us another piece of the puzzle.

Reading through volumes of scholarly studies on the origins of agriculture, one is struck by the diversity of opinion. Views about why agriculture began differ according to continent and region, and according to whether the researcher happens to be an anthropologist, archaeologist, plant geneticist, botanist, cytologist, historian, linguist, or meteorologist. Each has an answer, but none presents the entire picture. Together the pieces begin to make sense. Agriculture began between ten and fifteen thousand years ago through the efforts of hundreds of thousands of people on several continents, and in many different social and ecological situations. The agriculture they established, suited to their own needs, was developed over the course of several thousand years. Would it be reasonable to assume that there was a single cause, a single pathway that led to agriculture in each of these cases?

Initially agriculture developed in harmony with the needs and cycles of hunting and gathering, in order to supplement those food sources. But as agricultural pursuits became more productive and demanding, conflicts arose. In Central and South America, cultivation of corn and beans began to compete with the gathering of plants in the spring and fall, and with deer hunting in the rainy season.[37] Gradually hunting and gathering yielded to agriculture. And, as agriculture developed and dependence on it grew, the possibility of returning to a hunter-gatherer society diminished.

The transition from hunting and gathering to agriculture was rapid, given the span of human history, and it was not without its price. The Bible is replete with references to floods, droughts, and plagues. According to one theory, early agricultural practices were responsible for the destruction of the watershed of the Tigris River, which caused the great flood of biblical times, Noah's Flood.[38]

The Egyptian peasant might have been able to produce three times as much food as needed by the third millennium B.C.,[39] but an inscription on an Egyptian tomb dating from 2000 B.C. clearly addresses the price relatively stable hunting and gathering societies paid for their newfound dependence on agriculture.

> I kept alive Hefat and Hormer . . . at a time when . . . the land was in the wind and everyone was dying of hunger on this sandbank of hell. . . . All of Upper Egypt was dying of hunger to such a degree that everyone had come to eating his children. The entire country had become like a starved grasshopper . . .[40]

And from an eyewitness of the famines of the period comes this account of misery: "Hearts are violent, plague is throughout the land, blood is everywhere . . . many dead are buried in the river. . . . Indeed, the river is blood, yet men drink of it. . . . Why really crocodiles (sink) down because of what they have carried off, for men go to them of their own accord."[41]

FROM THE PLANTS' PERSPECTIVE

Primitive peoples encountered an overwhelming array of potential food sources in their environment before the days of agriculture. There were over two hundred thousand species of flowering plants alone from which to choose: thousands of fresh fruits, nuts, vegetables, and grains, although we would barely recognize some of them. Cauliflower simply did not exist. Beans existed, but the pods were no bigger than one's thumb and had miniscule seeds. Tomatoes were tiny. Most of the foods we cherish today bear little visual resemblance to their forebears eaten thousands of years ago.

Agriculture did not begin without some help, some adaptability, from the plants. Those first tentative moves that people made towards agriculture—first tolerating, then encouraging, and finally cultivating certain plants—produced startling changes in the plants themselves. The response of the plants to this attention gave people good reason to intensify their efforts. The process of domestication began to unfold—a process that continues today.

It was domestication—not civilization—that began with the sowing of seeds. Simple harvesting, which hunters and gatherers had practiced for millennia, initially resulted in few if any changes to the plants involved, for it was the seeds left unharvested that fell to the ground and produced the next generation. But hunters and gatherers eventually began the process of domestication with their increasingly intensive harvesting, and in due course their care and cultivation of select plants. With the advent of sowing, it was the harvested seeds that came to constitute the next generation. And those seeds changed history.

The simple act of harvesting seeds of nondomesticated plants and then sowing them produced remarkable changes of great advantage to people. Weeds and grass, as everyone knows, are extraordinarily adept at spreading their seeds. Walk through a field in early autumn and you will cause hundreds of seeds to fly in all directions. Turn your dog loose and it will often come back covered with seeds. Wild weeds and grasses are designed not to hold on to their seeds, but to shatter and drop them easily, even in the slightest of breezes. The survival of such plants depends on their ability to spread their seeds.

As hunters and gatherers walked through wild stands of wheat and barley, they could have harvested no more than half of the available seeds.[42] Most would have fallen to the ground. The seeds that remained on the stalk to be harvested often did so because of minor physical differences—differences not very conducive to a wild plant's survival, but most helpful to someone trying to collect the seeds without spreading them all over the ground. The unavoidable collection of non-shattering types caused the first fields planted by the first farmers to be constituted primarily of grasses significantly different in one respect from those that grew wild. Repeated sowings of these seeds produced non-shattering plants—plants whose seed or grain would remain on the plant even if jostled by the emerging farmer with a flint-bladed sickle. Genetically, the change was simple. Often the difference between shattering and non-shattering types is caused by just one or two genes, the biological bearers of heredity. With non-shattering grains, people were able to harvest a greater percentage of all the seeds in the field. Harvested yield increased, giving those first farmers positive response for their efforts.

In North America, interestingly, many grains were not domesticated—did not develop the non-shattering advantage—because of the way most Indians harvested the wild grasses. Typically, Indians beat the grain from grasses into baskets using small wooden paddles. Seed harvested by this ingenious method was seed that shattered easily, and retained its weedy characteristics. But this was neither the method nor the seed that would lead to domesticated cereal grains. (Some grains, like Sonoran panicgrass, were domesticated in North America, but are not in wide use today.)

To students of agricultural history, the non-shattering adaptation is the most striking and easily recognized trait of the domesticated crop. Domesticated crops cannot survive without cultivation because they have lost many survival mechanisms, notably efficient seed dispersal. But there are other differences between domesticated and wild plants. As "agriculture" began—even in the earliest and most modest forms of plant manipulation by hunting and gathering peoples—and as the process of domestication got under way, other modifications in the plants occurred almost incidentally.

Wild plants often have seeds that reach maturity over a long span of time. This is good for the wild plant, because it reduces the risk of some event destroying all the seeds at once. The seeds harvested and then sown by the first farmers were those that had matured at the same time—the day of the harvest. Thus plants with uniform maturing dates went on to constitute the next generation of plants in farmers' fields. Primitive sorghums and sunflowers had many seed heads to facilitate maturation over a period of time. The pressures of domestication, however, worked against this, so that today commercial varieties typically bear their seed in a single head. Domestication accounts for the difference one notices between the branching, multi-flowered wild sunflowers growing by the roadside and those tall, single-stalk, giant headed varieties tended in the garden.

Likewise, many wild plants have seeds that germinate irregularly. Some seeds will lie dormant in the soil while others from the same plant will sprout and grow. It is not nature's way to put all its eggs in one basket. Domesticated crops do not share this feature. Thousands of years ago those seeds that lay dormant and did not sprout with the others in farmers' fields did not mature in time to have their seeds harvested with the rest. Thus this dormancy trait was not carried forth in the seeds planted the following year. Moreover, seeds that sprouted first usually prospered most, getting a head start on other plants.[43] They produced many seeds that were harvested and resown, thus perpetuating this feature.

Domestication brought, and presumably still brings, even more changes. Large seeds provide more nourishment for the future plant. This means plants from large seeds have a competitive edge in crowded conditions. In the end they produce proportionally more healthy plants and more seeds than do plants grown from smaller seeds. Year by year they come to constitute a greater percentage of the crop. Plants with genes that give small seeds become rare. Over time this gradually increases the seed size of the crop.

In Mesoamerica, maize developed larger ears and larger kernels due to this process, and it began to ripen more uniformly. This, combined with the development of beans with pods that did not explode when ripe, made possible the corn and bean combination which remains the basis of South and Central American diets. It is probably no coincidence, as Frances Moore Lappé[44] aptly points out in *Diet for a Small Planet,* that this and other traditional diets include complementary proteins. They contain combinations of food tried and tested over thousands of years. And though first domesticated hundreds of miles apart from one another, wild beans and corn's wild relatives grew together in the same places. Harvesters could hardly encounter one without the other.

The loss of protective features of some plants—such as thorns, toxicity, and excessively fibrous tubers—would also have come about as a natural result of domestication, as people began to protect the plants from the predators for which these defenses had been developed.[45] The bitterness of cabbage, eggplants, and squash species disappeared, allowing more energy to go into increased production, and further enhancing the attractiveness of these plants to people. And plants developed responses to the types of agricultural practices employed. Were the fields weeded, watered, manured? If so, the plants adapted. If not, they adapted. Either way, the plants adjusted genetically to cultivation.

Many other changes took place in plants as a result of domestication. The changes listed above—non-shattering, increased seed size, uniform maturity, and loss of dormancy and defense mechanisms—were those that occurred almost solely due to the process of domestication. For the most part these adaptations would have been made with or without the awareness or conscious approval of the first farmers. They came as part of the package when people began to practice agriculture by sowing the seeds of their harvest.

The more plants altered in ways beneficial to mankind, the closer people were drawn to the demands and cycles of agriculture. Steadily the skills of hunting and gathering diminished or disappeared as agriculture progressed. The spread of agriculture, by gradually altering land use patterns and local ecologies, made it more and more difficult to return to hunting and gathering, for there was less and less to return to. Dependence on agriculture had its own way of fostering increased dependency.

In the process, it intensified people's relationship with the few plants undergoing domestication. Today, out of the thousands of plant foods once used by hunter-gatherers, only a small handful are employed. And of these, it is common to see just nine (wheat, rice, maize, barley, sorghum/millet, potato, sweet potato/yam, sugarcane, and soybean) mentioned as accounting for over three-fourths of the plant kingdom's contribution to human energy.[46] In some countries or regions other plants are important. But in total only about 130 species essentially feed us. Remarkably, after the initial agricultural impulse was over, our Stone Age ancestors had tamed and brought under cultivation virtually all of our major food crops.

Out of over a quarter-million flowering plants that exist, about 200–250 (excluding ornamental, pasture, and forest species) were domesticated.[47,48] It was an unprecedented achievement which we scarcely ever pause to appreciate.

Despite incredible advances in genetics and plant breeding, modern man has domesticated few—and some scientists would say no—major food crops. What we eat today we owe largely to our nameless ancestors and to a process begun in Neolithic times, long before recorded history. It is a process in which many unsung native peoples today are still engaged—the long process of domesticating plants.

Little in our eating habits remains to give evidence of our recent dependence on wild foods. Yet it was not until about the second century B.C., that cultivated plants constituted more than half of the human diet in such ten-thousand-year-old centers of agriculture as Central America.[49] Today wild rice features as a delicacy in many supermarkets, but one rarely encounters any other nondomesticated grain or vegetable in developed countries.

Some plants such as pigweed and lamb's quarters, once cultivated and perhaps domesticated in North America, have simply fallen out of use and are now only found as common weeds.[50] In West Africa, African rice is declining in use and is currently reverting to its wild state. Once a basic food in the diet, it is now considered a weed in the field of "Asian rice."[51] *Malva sylvestris* was already declining in use by the eleventh century and is now considered a weed.[52] *Sturtevant's Edible Plants of the World*[53] devotes five lines in 686 pages to this plant; *The Dictionary of Useful Plants*[54] by Coon makes only passing reference to its use in home remedies. Yet it was ancient China's most important green vegetable.

Wild foods are more commonly found today in Third World countries. A recent article about diet in Ethiopia indicates that war, tribal feuding, and natural disasters have renewed people's dependence on wild plants. The sustenance of some monks and nuns in isolated areas still comes largely from wild plants.[55] Mexico is attempting to stimulate a revival in nutritious wild plant foods. And in developed countries, increasing numbers of people collect wild, edible foods as a hobby.

But these are the exceptions. And while their importance is surely underestimated by the bias of our systems of collecting world agricultural statistics, nevertheless, barring a revival or a catastrophe, we are witnessing agriculture's final mop-up operation on hunting and gathering societies and on the plants they used.

C H A P T E R   T W O

Development of Diversity

Cauliflower is nothing but cabbage with a college education.—Mark Twain

"Pig droppings," Don Maximo yelled out with a big laugh. "We call it pig droppings. Just look at it." Robert Rhoades of the International Potato Center was sheepishly holding the long black object he had just dug out of Don Maximo's potato field high in the Andes. "What's wrong, amigo, don't you Americans know potatoes?"

Knowing potatoes means one thing to a New York shopper, quite another to a Peruvian peasant farmer. Of the five thousand potato varieties (modern commercial varieties and landraces) grown around the world today, Andean farmers cultivate some three thousand. Sometimes forty-five distinct varieties can be seen growing in a single field. They come in all shapes and sizes and a variety of colors including black, red, blue, purple, yellow, and white. And each variety has its own uniquely descriptive name, as Rhoades discovered.[1]

The twelve-thousand-year process of plant domestication produced an explosion of color and proliferation in the uses to which plants were put. When people began to domesticate plants, they did not do so for the purpose of creating non-shattering seed heads or seeds that were large and lacked dormancy. These were merely inevitable by-products of harvesting and sowing seed. People's interest in agriculture was in cultivating "crops" with certain characteristics of value, even if they did end up calling them "pig droppings."

People used sorghum as a grain, but they also wanted it for making molasses and brooms.[2] There were many species of gourds, valued for such things as food, musical instruments, utensils, and even as penis sheaths. Different types of maize were selected for flour, hominy, popping, boiling, for producing red-colored beverages, and for eating fresh off the cob. Red-kerneled maize was saved for use in ceremonies. When traits people wanted appeared, they were not allowed to be lost but were encouraged, maintained, and perpetuated by the acts of the first farmers. And those farmers' descendants all the way down to the present have seen to it that the traits have become integral parts of the crop itself. Indian corn is colorful not by accident, but because hundreds of generations have found pleasure and utility in the colorful ears. Blue and red pigments in cornstalks help corn varieties warm up quickly on cool mornings. Hence blue and red corns are often planted earlier than other varieties; the different colors in this case "mark" other associated traits.

Thus the process of domestication, of human selection, and encouragement of diverse plant characteristics continues even today. Jack Harlan of the University of Illinois once found an African farmer selecting crook-necked sorghum plants from his field to be saved as seed for next year's planting. Why did he save these types, Harlan inquired? Because, the farmer replied, they are easier to hang from the roof.[3]

Other farmers before him selected "sweet sorghum for chewing, white-seeded types for bread, small, dark red-seeded types for beer and strong-stemmed fibrous types for house construction and basketry."[4] The stems and leaves of some types were used as dyes.[5]

This process of selecting certain plants and sowing their seeds, repeated every year for thousands of years, can have effects which are a marvel to contemplate. The cultivated beet, a plant which originally resembled Swiss chard, was developed in a variety of directions.[6] Today some are leafy vegetables or salad plants. Some are root crops for people or animals. The sugar beet, developed from forage beets about two hundred years ago, provides much of the world's sugar.

Primitive cottons come in a number of fiber colors: white, brown, green and purplish-gray. Peasant farmers in Peru grow several different landraces on a small scale in order to fashion clothes with these natural pigmentations.[7,8]

Today, we appreciate the zinnia for its rainbow of colors: purple, red, yellow, orange, pink, white, green. But it was not always so. When the conquistadors marched through Mexico in 1519, they stepped right over the small purple and yellow flowers. The Aztec name for zinnias meant "eyesore," as it was used to cure ills of the eye. The Spanish followed suit calling it *mal de ojos*. But a German, Dr. Zinn, and later French horticulturists, saw other qualities in the weed and gave the world a favorite ornamental flower.[9]

And beans! What size, shape, or color do you desire? In Panama, people seem to prefer large red kidney beans; Venezuelans like black beans. East Africans are partial to mottled varieties.[10] John Withee, a retired medical photographer from Lynnfield, Massachusetts, likes them all. Some years back he set out to find the type of bean his parents had grown when he was a child, the Jacob's cattle bean. Withee is the persistent sort. He found his bean, but not until he had found and collected two hundred other varieties. Addicted, he continued collecting. People began to call him "the bean man." Today, Withee's living museum collection numbers well over a thousand varieties, few of which were ever found in a seed catalogue. He has constructed a display case which he carries about to fairs and garden club meetings to give people a sense of the diversity of beans. This diversity is the result of our ancestors appreciating variety (and, as we shall see in chapter eight, the efforts of people like Withee to save it).

People produced a wealth of crops from the cabbage species by selecting and encouraging the development of certain parts of the plant. Brussels sprouts, kohlrabi, cauliflower, broccoli, kale, and the vegetable we call cabbage all come from this species.[11]

With the development of each food crop, the impetus to domesticate others declined. Experiments continue and attempts at domestication persist even today—scores of ornamentals have been domesticated in this century. A number of *bona fide* crops are in the process of being domesticated by native peoples. And considerable attention is being paid by plant breeders to forages and forest trees. In terms of human history, the urge to domesticate what are now our major food plants was short-lived. After bringing several hundred plants under control, the epoch was over.

By cultivating certain species like beans year after year, our ancient hunter-gatherer then farmer ancestors were able to notice the mutations and changes that took place in their crops.[12] Favorable traits were selected and encouraged. In one village this might have meant that people developed a certain kind of bean, while people in another village nearby were busily encouraging production of an entirely different type. Mountainous areas were particularly conducive to enhancing diversity due to the obvious effect of diverse habitats. And the season when seeds would have been exchanged, facilitating the mixing of different types, was winter—when the mountains in some places were impassable.[13] Hence many different landraces of the same crop came to exist and persist.

The process of domestication forced plants to adapt to a different environment—the cultivated environment. In the wild, plants faced numerous ecological situations, but none quite like that provided by the first farmers. In these ancient gardens, the soil might be cultivated and weeds might be discouraged. The plants might be exposed to human and other wastes which made the soil more fertile. Drainage and irrigation were sometimes provided. Irrigation canals were being used more than seven thousand years ago in eastern Iraq, and drainage of swamplands for agricultural use was taking place in the highlands of Papua New Guinea five thousand years ago.[14] These practices reduced and simplified the diversity of the natural world. But they created no utopia for the young crops. The new crops were forced to adapt to conditions of life under primitive cultivation. They found their niche in the world by surviving under conditions of low fertility, environmental fluctuations, varying soil types, and subjection to pests and diseases. Differing conditions between one village and the next resulted in genetically different varieties being developed.

THE BIRDS AND THE BEES . . .

In plants, the shape and color of the leaves, the nature and extent of disease and pest resistance, the ability of the plant to survive in arid conditions or flourish in tropical forests—these and countless other attributes are determined by the plant's genetic makeup. A plant cell may contain a hundred thousand genes. Sometimes it is a single gene that alone produces a certain trait. The non-shattering quality of domesticated grains is often the result of just a one or two gene difference between the wild and domesticated types. Other qualities like pest resistance are more often the result of many genes working "in coordination" with each other.

When two plants interbreed or "cross," their offspring typically receive a combination of genes—half from each parent. With this new assemblage of genes, we can expect the offspring to be adapted to those conditions to which their parents are adapted, or to intermediate conditions. The genes and the way in which those genes are configured in a cell ultimately direct the production of chemicals which determine the plant's characteristics.

More often than not, it is two similar plants from the same locality that interbreed, producing offspring only slightly different from themselves. It is the offspring most closely resembling the parents which are likely to survive, for they will be adapted to the existing conditions.

Until recently evolution has generally been viewed as a long, slow, barely perceptible process involving natural selection acting on occasional random mutations. But we now know that plant populations have the ability to make rapid changes through a number of genetic mechanisms whereby they can adapt to the world of myriad and changing combinations of soils, climates, pests, and other factors. These mechanisms even account for the existence of many weeds, some of which are nothing more complicated than crosses between wild and domesticated plants. It stands to reason: weeds have some of the characteristics of both. They are adapted to cultivated (disturbed) areas, and, as with domesticated plants, few could flourish without people; but are they wild! All cultivated cereals have companion weed forms. There are also weed carrots, weed watermelons, peppers, potatoes, sunflowers, and others.[15] Let it be noted, however, that a number of these weeds are crop progenitors, and not the offspring of crop and wild plant crosses.

Plants have built-in genetic barriers against certain crosses. The two plants breeding must be compatible in basic ways. One will never (let us hope) see bananas and eggplants interbreeding, but domesticated and wild lettuce will cross. Studies in Pullman, Washington, indicated that this often happens unbeknownst to local gardeners, because the offspring so closely resemble the wild type.[16] And in the Nobogame Valley of Mexico, corn occasionally hybridizes with its primitive ancestor *teosinte* volunteering in fields. There, farmers believe that the *teosinte*'s presence with corn enables the next corn planting to produce better crops.[17] The Tsembaga people in New Guinea follow a similar practice with hogs. They castrate the male pigs and allow the sows to mate with feral boars in the forest in order to improve hardiness and disease resistance.[18]

From the early days of plant life on earth, the ranks of flowering plants have grown steadily.[19] Wind and insects are the major pollinators of these plants. Insects like bees are good at searching out isolated flowers, so that plants from the meadow are often crossed with those growing in the woodland. The 334 species of insects that visited carrot flowers in a study in Utah would surely provide ample opportunity for the carrots to breed with other carrots or with related wild and weedy species.[20] Perhaps flowering plants have flourished precisely because they had a breeding system that encouraged the creation of offspring adapted to so many different environments.[21]

Adaptations can be so sensitive. In a five-meter area, Dr. R.W. Allard of the University of California found striking genetic differences between flowering times in wild oats, apparently associated with local topography. Flowering time varied by more than two weeks in wild oats growing from the top to the bottom of a steep slope. Normally such a difference would only occur over a north-south distance of eight hundred kilometers.[22] Scientists studying grass growing on the edge of a zinc mine found that the grass in soils with high zinc content was tolerant to the metal, while grass just three meters away showed little tolerance.[23] As one might expect, when left alone without the benefit of modern plant breeding, plants adapt themselves quite admirably. In northern Nigeria for example, the flowering dates of local sorghum correlate closely with the average date of the close of the rainy season.[24] In India some wheat varieties have developed a novel approach to extreme drought conditions: they shed lower leaves to form mulch, which helps retain soil moisture.[25]

With the birth of agriculture, new environments were being created. People were scratching at the earth. They fertilized, irrigated, and drained it. They cut trees and burned meadows. And when they moved, ancient farmers carried the seeds into new habitats where they mingled with the native plants to produce new genetic combinations—new plants. The greatest plant breeding experiment in history was just beginning twelve thousand years ago. "In the millennia-long melting pot of plant races thus created, genetic exchange between plant races and species took place on a scale never before—or since—possible," says plant geneticist Erna Bennett, who describes the resulting creation of new plant forms as an "explosion . . . a flood of evolution."[26] The diversity thus released enabled plants to become adapted to minute changes in the environment.

Just as crops and other plants became adapted to soil conditions, rainfall, the length of daylight and the like, so they also evolved with insects and diseases. Over 260 diseases and pests strike the potato.[27] But along the way the potato developed defenses against them. Had it not, would we even know what a potato was? One potato species with a unique approach traps insects with a sticky secretion from its hairy leaves.[28]

The key to understanding pest and disease resistance in plants is to remember that plants and pests *coevolved*. When the plant developed defenses, the pest or pathogen adapted with new methods of attack. An imprisoned boll weevil found in a cotton boll dating back to 700 A.D. shows what old foes these two are.[29] But the continued existence of ancient cotton varieties shows that cotton developed its defenses.

In those cases where plants did not evolve in the presence of a certain disease, they have been forced to develop that resistance later (through introgression or other means) when disease met plant. Upland cotton picked up resistance to black-arm from native or wild species of cotton when introduced into Africa, for example.[30]

DIVERSITY OF DIVERSITY

During the thousands of years that most crops have been under domestication, they have encountered and been forced to adapt to almost every conceivable condition. Consider that apricots, a warm climate fruit, are grown in the Himalayas where night temperatures persistently fall below freezing.[31] Sorghum grows from the wet tropics of West Africa to Asia's semi-arid regions.[32] In India, rice grows from sea level to 7,000 feet. Some varieties grow under fifteen to twenty feet of water. Others are adapted to grow in areas with as little as twenty-five to thirty inches of rainfall annually.[33] Potatoes grow from below sea level to 14,000 feet, from the Arctic Circle to southern Africa.[34]

Not all sorghum, rice, and potato varieties grow in such diverse climates. But many crops have developed genetically distinct varieties that are adapted to such a range. And at each step along the way of that range, crops were forced to adapt to the conditions at hand—adapt or perish. Whether by mutation or introgression (the introduction of a gene from one gene complex to another, through cross-breeding, for example), the crops did just this. The crops that peasants grow today—direct descendants of the crops our ancestors began to domesticate twelve thousand years ago—exist for a good reason. They are well adapted and well endowed with genetic variability. With these characteristics, the result of thousands of years of encounters with pests and diseases and changing environments, they are able to continue to evolve and adapt to new conditions.

No one knows exactly how many genetically distinct varieties of any crop were created by natural and human selection. No one knows for sure how many exist today. Few have even ventured a guess. Suffice it to say that tens of thousands of genetically distinct varieties of crops like wheat, corn, and rice exist.

In the United States alone, more than seven thousand named varieties of apples and over twenty-five hundred kinds of pears are known to have been grown in the last century.[35]

At an exhibition held at the Crystal Palace in London in 1900, on the occasion of the two hundredth anniversary of the introduction of sweet peas to England, 264 varieties were shown.[36] That is an impressive number, but surely many more existed. At the same time, fifteen hundred varieties of fuchsias were known in England.[37]

This diversity can be illustrated in yet another way. Consider corn. A Papago Indian type in North America matures as a dry flour corn in fifty-five days, while one Colombian variety requires sixteen months. Some corns have just eight leaves, others have up to forty-two. Plant height varies from 40 cm to 700 cm; ear length ranges from four to 40 cm; and the number of rows of kernels from eight to twenty-six. The weight of a thousand kernels can be as little as 50 grams for one Peruvian variety to as much as 1,200 grams for another Peruvian variety.[38]

Squash can range from the size of a chicken's egg to over a half-meter in diameter. Weight and size within fruits and root crops can vary a thousandfold.[39] Some radishes could easily hide in a small salad, while a single radish of another variety might weigh sixty-five pounds and fill a bucket.[40]

Diversity appears in the different colors of flowers and beans and cotton lint. It is what enables crops to grow in such different environments. It provides resistance to pests and diseases. We see it when we examine the height of a corn plant. We even taste diversity. Tomatoes, peppers, corn, and other crops offer a wide range of tastes among varieties.

The existence of such diversity has been recognized for thousands of years. In 300 B.C., Plato's pupil Theophrastus wrote in his *Enquiry into Plants* of the many types of wheat which differ in "color, size, form, and individual character, and also as regards their capacities in general and especially their value as food."[41]

The diversity Theophrastus noticed and which we still see today was not created overnight, nor did it always exist. A few crops are of relatively recent origin. Brussels sprouts originated in Belgium in the eighteenth century and were produced from the cabbage species.[42] Pyrethrum, a flower now used as the active ingredient in some "natural" pesticides, was established as a crop in the first half of the twentieth century on the Dalmatian coast of Yugoslavia.[43]

But the great majority of our crops are of ancient origin, and the diverse varieties of each crop owe their existence to thousands of years of evolution under domestication. It was the result of this long evolution and the influence of people in that process that Darwin noticed. Significantly, he titled the first chapter of *The Origin of Species*: "Variation Under Domestication."

Darwin could not have known about the origins of recent crops like pyrethrum. But he realized that people had a hand in creating diversity—it was not just visited upon them. Varieties adapted to and depending for their survival on irrigation, for example, would not have existed before that practice. As simple as these observations may seem, they were necessary in order for the next big breakthrough in our knowledge of crop diversity to take place. This breakthrough came in the Soviet Union.

GEOGRAPHY OF DIVERSITY

In 1929, Nikolai Ivanovich Vavilov was in the middle of his career as one of the most outstanding biologists, geneticists, and plant explorers of the century. He was in Japan, having travelled there to collect wheat specimens. Full of excitement, he climbed up the steep little steps into his railcar at the Kyoto railway station and shouted goodbye to his friends: "Sakurajima-Daikon, Sakurajima-Daikon!" Sakurajima-Daikon is not the Japanese word for goodbye. It is the name of an exceptionally large radish.[44]

Eleven years later his remarkable career was tragically ended. What follows is part of his story.

Vavilov was born into a wealthy merchant family, destined, it seemed, to produce several outstanding scientists. By all accounts Nikolai matured into a kind and gentle man—a man of immense talents, driven by his science.

Vavilov's memory was legendary. He could recite books by Pushkin, word for word, from memory. He knew English, German, Latin, French, Spanish, Farsi, and Turkic. He found little time for rest. The only time he was seen to sleep in public was during a particularly turbulent flight with a group of scientists en route to Baku on the Caspian Sea. Some in the party were busily scratching out their wills, while others were too scared even to do that. Failing to interest any in conversation and seeing no way to advance science, Vavilov fell asleep.

"Life is short, we must hurry," he often said. And hurry he did. His first expedition abroad solely to collect plants was to Iran in 1916. While there he was mobbed, attacked, and then deserted by his guides. Returning to Russia with his German textbooks and English notes, he was promptly arrested at the border as a spy. Three days later he and the samples were released. Those samples would found the world's largest seed collection.

His collecting expeditions brought him encounters with malaria in Syria, typhus and bandits in Ethiopia, a landslide in the Caucasus Mountains, and a plane crash followed by a sleepless night next to a lion's den in the Sahara. He even smuggled some American guayule (a rubber-producing plant) back to the Soviet Union. These travels also brought him into contact with more crop diversity than anyone had seen before. And they brought the Soviet Union immense variability from which to fashion new crop varieties for its agriculture. But on a collecting expedition to the western Ukraine, Vavilov's story was to take a dramatic turn. He was arrested.

On August 6, 1940, in Chernovicy near the Romanian border, a big black car occupied by agents pulled up. Vavilov's colleagues were told he was needed in Moscow for urgent consultations. The consultations turned out to be questionings. He was under arrest. Subjected to four hundred interrogations lasting for seventeen hundred hours over an eleven-month period, Vavilov finally confessed to high crimes. In a five-minute trial without lawyers, he renounced the confessions but was nevertheless found guilty of "belonging to a rightest conspiracy, spying for England, and sabotage of agriculture."[45,46]

He was imprisoned in Moscow and later transferred to a prison camp in Saratov, where he had begun his career as a teacher at the university. For two years he awaited execution. During this time it is thought that he wrote a long work entitled "A History of the Development of Agriculture," but the manuscript was never found. Meanwhile his friends and family lobbied for his release (several were fired or jailed for their efforts), though they had no way of knowing if he were even alive. His death sentence was commuted in the summer of 1942, but he was never released. On January 26, 1943, he died in a prison hospital in Saratov. A Soviet journalist authorized in the 1960s to look into Vavilov's death apparently uncovered autopsy records indicating that Vavilov died of starvation.[47]

Vavilov's mistake was being a good scientist during times in which being a good scientist was bad politics. During the 1930s in the USSR, genetics came under extreme attack from the Michurinist movement led by T.D. Lysenko, a Soviet biologist (who was to replace Vavilov as the director of the Institute of Genetics). In the U.S., ironically, genetics is now coming under attack by the Moral Majority and other right-wingers. They charge that it is "secular humanism" and anti-Christian. Backed by Stalin and Moscow, Lysenko charged that it was fascist and anti-Communist.[48,49]

Lysenko believed in the inheritance of acquired characteristics. He believed that genes could be "trained." He believed and claimed to have proven that spring wheat varieties could be coaxed into becoming winter wheat varieties and *vice versa* simply by exposing the seed to soaking in water of different temperatures. Vavilov, however, was a scientist. His own knowledge and study of genetics convinced him that Lysenko was quite wrong.

Vavilov was the director of the All Union Institute of Applied Botany and New Crops, a collection of four hundred research institutes with twenty thousand employees. He was a member of the Soviet Central Executive Committee, past president of the Academy of Sciences of the USSR, and president-elect of the International Conference of Genetics. He was among the first recipients of the prestigious Lenin Prize. In short, Vavilov's influence was considerable and well deserved.

Lysenko and his followers were agitating to remove genetics from the universities. The political tension surrounding the study of genetics was reaching crisis point. Back in the U.S. a young Jack Harlan had convinced his father, a noted agronomist, to write and inquire about the possibility of the young Harlan going to the USSR to study under Vavilov. Vavilov was close to the elder Harlan and had stayed in the Harlan home on his travels to the U.S. It was late in 1938.

Vavilov's reply came quickly. The elder Harlan opened the letter and began to read something about agriculture in China. It was an elaborate code agreed on between the two old friends in 1932 at the International Genetics Congress at Ithaca. The elder Harlan read aloud to his son, and became more and more agitated: "Vavilov says you must not come. You could do no useful work. You would even be in danger yourself. Vavilov says that he fights with his back against the wall. Vavilov says he will never surrender."

Dr. Harry V. Harlan closed the letter and turned toward his son. Dr. Jack Harlan remembers to this day the sadness in his father's eyes and the remark, "Things must be truly terrible for him there now."[50]

Lysenko's brand of science had gained Stalin's blessings. It emphasized that traits acquired during life could be passed on to the next generation. Lysenko's theories seemed to Stalin consistent with his view of historical materialism, and they promised quick gains for Soviet agriculture, to boot. Vavilov was in their way; he had to go. His arrest capped a period of growing ridicule and opposition to his theories of genetics inside the Soviet Union, and growing respect and acceptance of those theories everywhere else. But Lysenko's views were politically acceptable. Vavilov's arrest did not come without warning. At least eighteen others in his institute had already been arrested and carried off. Though Vavilov's "Work Plan for 1940–41" taken from his notebook revealed that he planned to write twelve books and five major articles during that period, surely he must have known what was coming. Perhaps he just chose to work until they came to get him.

In spite of official support, opposition to Lysenko by Soviet scientists existed and survived. His attempt to induct a crony into the Academy of Sciences of USSR was openly denounced in a stunning statement from physicist Andrei Sakharov. A vote was taken. Lysenko lost. Khrushchev was enraged and considered disbanding the academy. But he had other problems. So did Lysenko; he stepped down for the last time as president of the Academy of Sciences in 1962, after a period of almost ten years of growing criticism.

On October 13, 1964, a Soviet geneticist was given twenty-four hours to prepare "a full-dress report on the achievement of genetics" for the Central Executive Committee. The next day Khrushchev was ousted. Lysenkoism was dead, the fraud of his "science" finally exposed.[51] Genetics was restored.

Four years later the institute Vavilov had headed was renamed in his honor in time for its seventy-fifth anniversary.[52] A postage stamp was even issued in his honor.

It is a painful history. But years before, as Vavilov's train pulled out of the Kyoto station on that day in 1929, he must have been bubbling over. On his travels, he was seeing things no other geneticist or biologist had ever seen before—the marvelous diversity not just of radishes in Japan, but wheats in the Near East and potatoes in the Andes. In his time, Vavilov was the world's most widely travelled biologist. His plant collecting expeditions had taken him to sixty-four countries.[53]

In the two decades that Vavilov was scouring the countryside, his teams added a quarter of a million entries to Soviet seed collections. No country since has come close to duplicating this feat. He combed the Middle East,

Afghanistan, and Ethiopia for primitive wheat varieties. He journeyed to North and South America and to the Far East. But more importantly than travelling and collecting widely, he began to notice a pattern.

Genetic variation—the diversity created by thousands of years of agriculture—was not equally distributed around the globe. In a small, isolated pocket on the Ethiopian plateau, Vavilov found hundreds of endemic varieties of ancient wheat. Studying other crops, he found some regions blessed with astonishing diversity, while other areas were relatively impoverished. In the following years, observations by other scientists confirmed Vavilov's budding theory. While living in a suburb of Guadalajara, Mexico, Edgar Anderson noted that he found "more variation in the corn of this one little township than in all of the maize in the United States."[54]

Vavilov mapped out the distribution of this diversity for each of the crops he studied. He reasoned that the degree of diversity was indicative of how long the crop had been grown in that area. The longer the crop had been grown, the more diversity it would display. More uses of the crop would have developed and this would be reflected in the variety of forms: corn for popping, corn for ceremonial and medicinal purposes, corn for roasting, and so on. More colors and textures could have been selected. More mechanisms for pest and disease resistance could have evolved. A recent arrival, on the other hand, would not have had a chance to produce many different varieties.

By locating a center of genetic diversity for a crop, one pinpointed its origin, Vavilov reasoned. This was where the crop had originated and had had time and opportunity to develop wide diversity. A plant's "center of diversity" was thus its "center of origin," he said. (Vavilov's "centers of origin" are shown in Map 1.) In the Middle East one found the wild plant relatives of wheat and the greatest diversity of primitive wheats. It was here that wheat had originated and not in the United States, for example, where one found none of this.

With this insight, Vavilov was able to look back through the darkness of ancient history. Conventional wisdom had assumed that agriculture had arisen along the Tigris and Euphrates rivers. Vavilov was discovering otherwise. Diversity was concentrated "in the strip between 20 degrees and 45 degrees north latitude, near the higher mountain ranges, the Himalayas, the Hindu Kush, those of the Near East, the Balkans, and the Apennines. In the Old World this strip follows the latitudes, while in the New World it runs longitudinally, in both cases conforming to the general direction of the great mountain ranges."[55] The mountains provided ideal conditions for the rise of diversity: varied topography, soil types, and climates. And they were excellent barriers to outside incursions and even local exchange, thereby sheltering their diversity.

Vavilov also hypothesized that these "mountainous districts are not only centers of the diversity of cultivated plants but also of the diversity of human tribes."[56] As we have seen, human cultures played a major role in creating diversity in agricultural crops. Vavilov found that ethnographic maps of his centers of origin were good indicators of where plant diversity would be found.

As Vavilov discovered what he thought to be the centers of origin for more and more crops, he noticed that they overlapped. The center for wheat is not the center of origin for wheat alone, for here a great diversity of barley, rye, lentils, figs, peas, flax, and other crops is also found.[57] These crops share a common center of origin.

Thus, Vavilov theorized that the world's crops had originated in eight definable centers of origin. It was in these centers—all located in Third World countries—that agriculture had originated, he suggested, and that the greatest genetic diversity was to be found. The eight centers (Map 1) were listed as follows: China; India, with a related center in Indo-Malaya; Central Asia; the Near East; the Mediterranean; Abyssinia (Ethiopia); southern Mexico and Central America; and South America (Peru, Ecuador, and Bolivia), with two lesser centers—the island of Chiloe off the coast of southern Chile, and an eastern secondary center in Brazil and Paraguay.[58] For each center he listed the crops he thought had originated there: 136 crops in the Chinese center, 84 in the Mediterranean, and so forth. Each of the centers developed both cereal and legume plant sources, making for balanced protein in the diet.[59]

Researchers have subsequently altered Vavilov's eight original centers by re-drawing their boundaries to reflect growing knowledge of agricultural origins and diversity. In the process, the number and the size of centers have also increased (see Map 2 for an example of a new formulation). Some workers now seem to have settled on twelve centers of diversity/origin, not all of which lie completely within the Third World.

In the last forty years however, new techniques and entire scientific disciplines have provided a more complete view of crop origins. History, religion, anthropology, cytology, archaeology, and even linguistics all play a role in the study of plant origins. Linguists will note that the Chinese characters for wheat and barley are derived from the character *lai,* which means "come." The *Chinese Book of Odes,* which dates from the sixth to

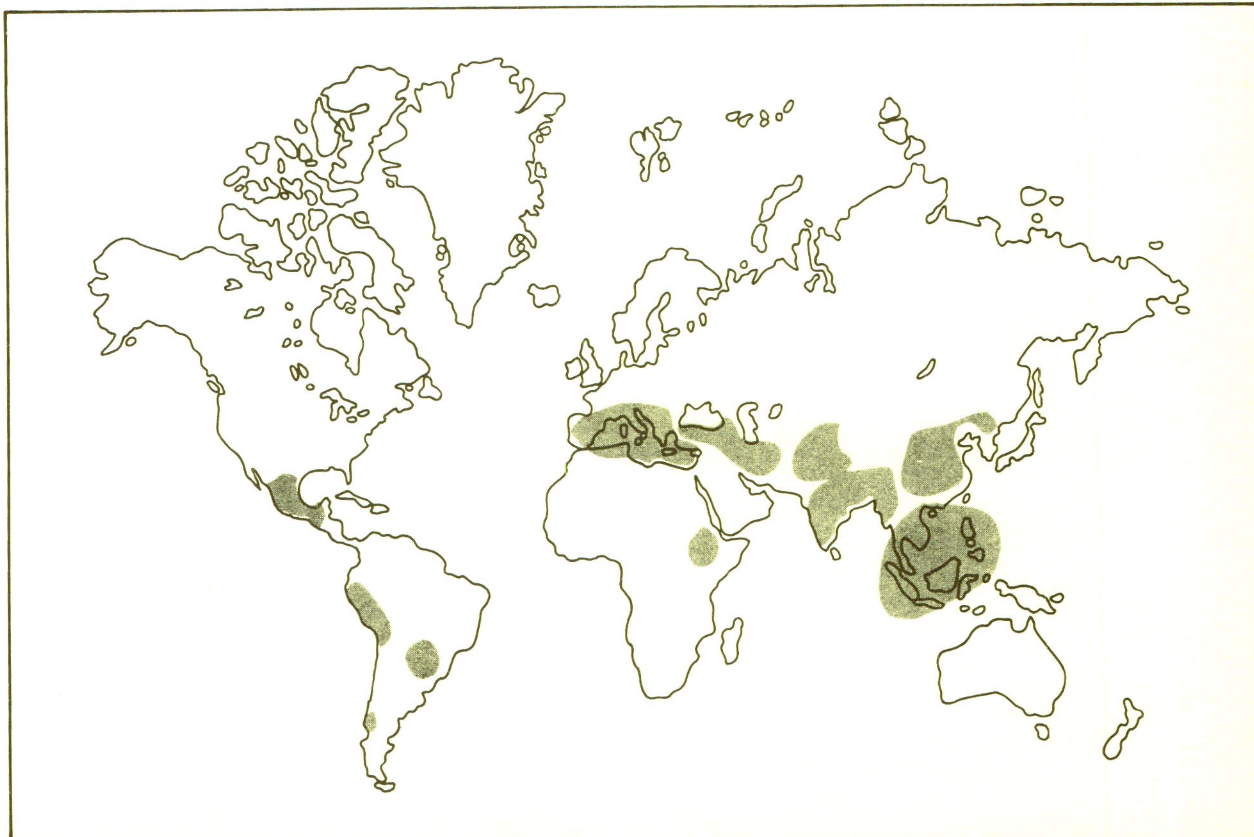

Map 1.
The centers of origin of cultivated plants according to Vavilov.

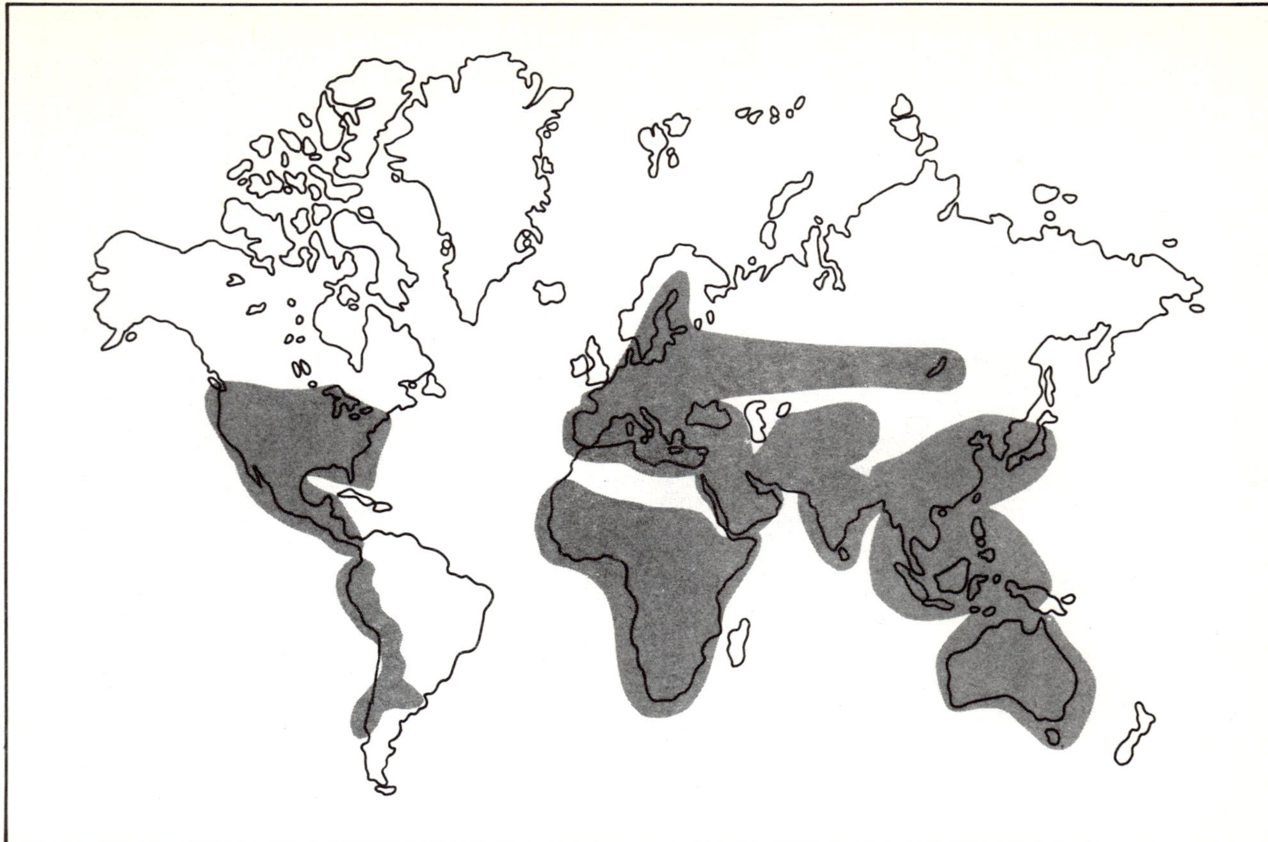

Map 2.
The enlarged centers of diversity from Zhukovskij based on the original Vavilov theory.

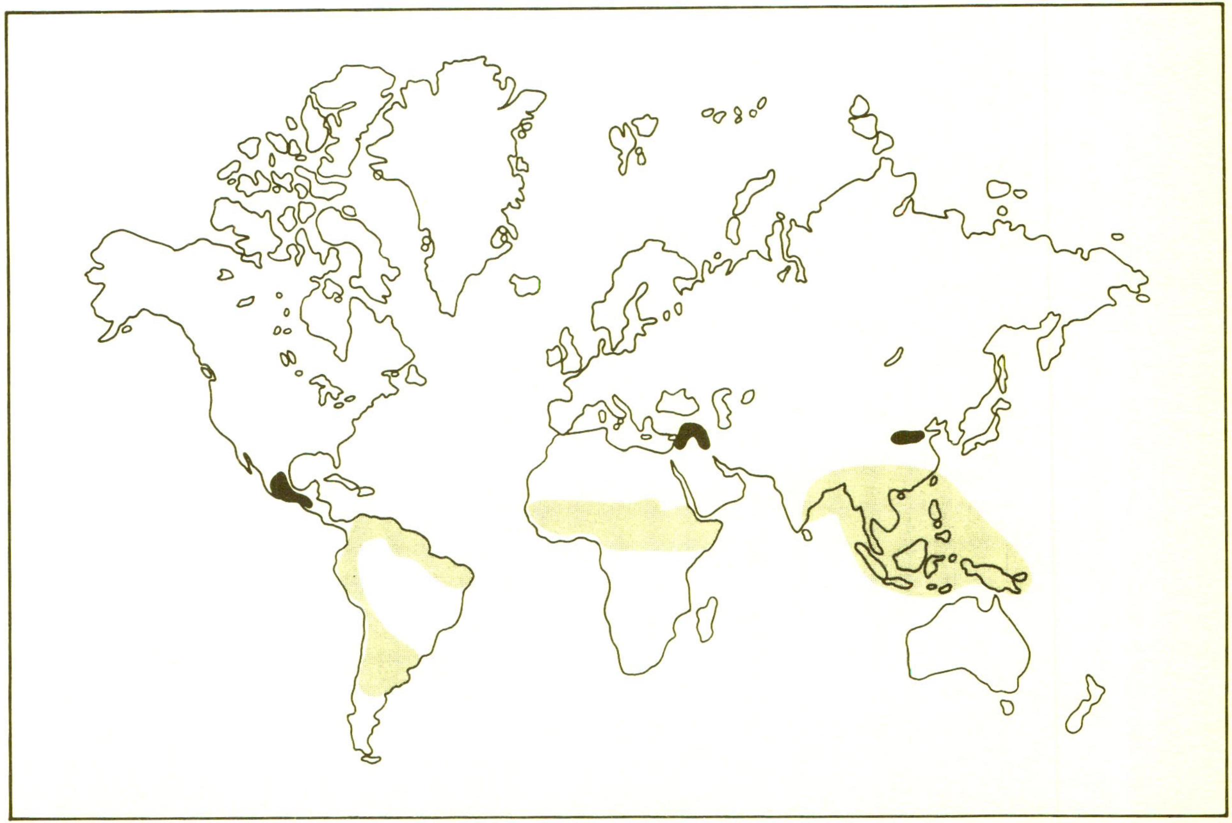

Map 3.
The centers and non-centers of origin from Jack Harlan.
Centers are shaded. The larger non-centers are outlined.

the eleventh centuries B.C., mentions the native origins of some crops, but never fails to point out that wheat and barley were given to the people by the "Supreme Ancestor." The authors evidently knew the two were not native and had no other way to account for them; this appears to indicate that wheat and barley were not native to China, but were introduced in very early times.[60]

Further studies of the distribution of diversity and the presence of wild crop relatives, the use of carbon 14 dating techniques, and the results coming out of various archaeological excavations have enabled contemporary scientists to challenge and amend parts of the theory.

Most now accept the useful concept of centers of diversity. That centers of diversity were centers of origin was Vavilov's assumption. Late in his career, Vavilov realized that the two are not always the same. Centers of diversity do not always comprise centers of origin; crops may originate in one place and yet develop much of their diversity in another. Despite tremendous genetic diversity in Ethiopia, it is doubtful that wheat originated there, for none of its primitive ancestors can be found growing there. Perhaps, then, this diversity is due to wheat's early introduction to Ethiopia and its long cultivation by many cultures. Similarly, scientists have considered the tomato to have been domesticated in Mexico and Central America, where most cultivated tomatoes originated. Yet this crop is descended from a wild species indigenous to the Andes, and more wild species occur there.[61]

Other crops like sorghum were domesticated over a broad area. The common bean was probably domesticated independently in Mexico and further south. A single, specific, clearly defined center does not exist.

Finally, to draw a circle around a continent, as some contemporary geneticists have done, in order to encompass the centers of origin of all the individual crops, and then to call it a "center of origin" of agriculture is to distort the meaning of the word "center" and render the whole concept hopelessly vague.

What has emerged in the 1980s is this: certain well-defined centers in the Vavilovian sense do exist—in the Near East, in northern China, and in Mesoamerica (Map 3). Here we have a reasonably clear picture of ancient agriculture emerging independently and a whole complex of primitive crops being domesticated. These crops then spread outside the centers, some establishing "secondary centers of diversity." Many of the Near Eastern crops—barley, flax, oats, peas, and the various wheats—have such a secondary center in Ethiopia.

Dr. Harlan speaks of "non-centers" in Africa, southeast Asia and the South Pacific, and in South America, as being loosely associated with the more defined centers.[62] Domestication has also taken place in these non-centers and great diversity is found there, but it is harder to argue for the existence of a clearly delineated "center" in any meaningful sense of the word. Harlan's theory leaves many questions unanswered, but it fits more comfortably with reality. (Harlan's concept of centers and non-centers is shown in Map 3.) While it is possible to talk about centers of origin and diversity for agriculture as a whole, there is usually more accuracy and utility in considering the centers of diversity of individual crops.

Vavilov correctly assumed that genes for disease and pest resistance would be found in the centers, because it was there that plant and pest had coevolved the longest. Vavilov even observed that different plants developed similar attributes in the same environment, and developed a theory from this observation. The practical significance of this is tremendous. If a plant breeder seeks genes for disease resistance in potatoes, it is good to know that the Andes will be a likely place to find them. If a certain resistance in one cereal is found in Afghanistan, it helps to know that other cereals in Afghanistan will probably also display resistance to that disease. Otherwise, the world can be an awfully big place to scour.

The beauty, simplicity, and utility of Vavilov's theory of centers remain despite the enlargements and modifications that have been made by Harlan and others. If it does not always make sense to speak of centers of origin as Vavilov did, it is still essential to understand that crops have centers of diversity. (And for the crop evolutionist, this diversity remains a crucial clue in delving into the crop's origin.)

It is crucial to understand that plant diversity is not spread evenly around the globe. In the future, it will be important for us in developed countries to understand that our agriculture is imported. We do not intend to deny the importance of secondary centers of diversity. The U.S. is such a center for lima beans, corn, pumpkins and squash—crops grown by Indians. We wish, however, to stress our debt to other regions, for all of the major, staple crops we grow and eat have their origins and their primary centers of diversity in the Third World. Americans settling down to a Thanksgiving dinner of native crops still commercially grown would be staring at a plate of blackberries, cranberries, sunflower seeds, and Jerusalem artichokes. (Ah, and there would be turkey!) But pity the poor Australians: their only native, edible plant grown as a crop, the Macadamia nut, was domesticated in Hawaii.[63]

PLANTS AS HITCHHIKERS

As long as diversity was being created and maintained by tribes or ethnic groups, crops spread only slowly outside their original areas of domestication. Ironically, the failure of the first agricultural systems may have led to their spread. David Rindos and others have argued that when early agriculture-dependent people experienced a crop failure, they had two options. Some could leave in search of better conditions, taking their agricultural techniques and crops with them. Others could stay and intensify their efforts. If successful, the human population would be able to grow, once again putting stress on the already precarious agro-ecosytem. The system that was most efficient at spreading would be the one subject to frequent, but not devastating, failure. Increasing dependence on agriculture resulted in failures of the system, which in turn resulted in the system's expansion with the spread of its crops.[64]

Still, it was not until the seventh century A.D. that wheat reached Japan by way of China, for example.[65] And soybeans from China did not reach Europe and the U.S. before the eighteenth century.[66] Tribal boundaries, topographical factors and lack of transportation systems combined to impede the spread of a crop and its diversity from one place to another. But with the breakdown of these barriers, largely due to war and trade, plants began to move as never before.[67]

Crops spread until they encountered either their agricultural/physical limits or competition from rival crops that were more economical to grow. To generalize, the longitudinal (east-west) spread of crops in ancient times had to do with historical factors, often dependent on race and culture. In moving longitudinally the crops encountered more cultural than botanical barriers, though some crops were enthusiastically accepted by most cultures.

The high latitudinal (north-south) movement of crops was another matter. Generalizing more dangerously, northern boundaries in the northern hemisphere (particularly in the Old World) and southern boundaries in the southern hemisphere were often agricultural ones. Few crops originated in these high latitudes to compete with other crops moving in that direction. Crops ceased to spread when they encountered weather that was too cold, daylight hours that were too short or long, or other similar factors. The low latitudinal limits were generally reached when the plant approached the tropics and came into competition with a more valuable tropical plant. The strawberry, for example, has not always been able to compete with tropical fruit crops in the low latitudes.[68] There are too many tropical fruits more productive there than the strawberry.

Crops of tropical origin have generally remained tropical, but there are interesting exceptions. A number of important crops which are perennials and grow year-round, year after year, in their tropical homelands, are cultivated as annuals in the higher latitudes. These include cotton, castor bean, peanut, and tobacco.

Until seven hundred years ago, all cottons in India were perennial shrubs susceptible to frost damage. Like other perennials-turned-annuals, they were selected (by the Aztecs, Arab peoples, and others) for forms that fruited quickly enough to produce some yield in the first growing season.[69] These forms could then be used as annuals where winter frosts would have killed the perennials. And in time the yields could be increased, for annuals tend to put more energy into the fruit and less into vegetative growth than do perennials.[70]

Other tropical plants were able to spread far beyond their homeland by finding a new home in the long, hot summers of temperate regions. Corn and the common bean are examples.[71]

The migration of some crops helped to bring others into cultivation. As wheat spread northward it was accompanied by weedy rye which infested the wheat fields. Rye was more tolerant of cold than wheat, so as wheat began to reach its northern limits, ancient farmers must have noticed the potential value of the weed, rye. At this "border," it has been customary for farmers to sow both wheat and rye. If the wheat does not make it, the rye will. Farther north, rye became predominant as the valued grain.[72]

Crops spread from their original area of domestication in every way imaginable. Like all emigrants, early farmers took seeds with them. Some crops did not need much help. It is thought that gourds of African origin "could easily have been dispersed from Africa to the New World by ocean currents" as early as 5000 B.C.[73] The bottle gourd may be the oldest annual cultivated plant on several continents.

Ancient rock drawings in the southwestern United States frequently depict the legendary hunchbacked flute player Kokopeli. It is said that Kokopeli did not actually have a hunchback, but that he distributed gifts from a basket on his back, and the basket gave him a hunchbacked appearance in profile on a cave wall. Today, flute playing medicine men of the Callalhuayo Indians of the Andes travel into Central America with packs on their backs. One of the items in each pack is the ancient and very odd "podded" corn. Was the legendary Kokopeli an ancestor of these medicine men? Were medicine men responsible for introducing podded corn as far north as Utah, a continent away from its home in South America?[74] A Mesoamerican deity of fertility is named Xochipilli; there seems little doubt that seed trade and exchange did occur between Mesoamerica and what is now the southwestern United States.

As crops spread, they encountered new peoples, new conditions, and new environments. This generated more diversity as the new crops crossed with other crop varieties, weeds, and wild relatives. We would not recognize many of the crops as they looked at the time they spread past their original homelands. It was only later, after they had developed additional traits, bred with their wild and weedy relatives, and been subjected to further human selection, that they came to look like the crops we know today. Thus the creation of the crops was a process, not an event that can be tied to one locality.[75]

Moving into recorded history, it is Egypt's Queen Hatshepsut who deserves credit for dispatching the first known plant hunting expedition, about 1482 B.C.[76]

Europeans dominated plant explorations from the sixteenth to the nineteenth centuries. Their legendary expeditions are well known. In the motherlands of Europe, plant collecting and the cultivation of exotic flowers from the New World became both fashionable and profitable. In the U.S., government officials travelling abroad were asked to send back useful plants. Thomas Jefferson smuggled upland rice out of Italy in his coat to introduce it to South Carolina, though the farmers regarded it as inferior to their own wetland varieties and begged him to keep his rice to himself.[77]

With the rise of colonialism, crops were introduced into new regions to be grown on plantations tended by people either formally or informally enslaved.

One well-known story bears repeating. A group of planters in the West Indies petitioned King George III of England to help them obtain breadfruit from Polynesia to help feed their slaves. He agreed and in December, 1787, William Bligh left on the *Bounty* for Tahiti. After propagating the breadfruit there, he set sail for the West Indies in April, 1789, with over a thousand breadfruit and other plants in hundreds of pots and tubs. Scarcely three weeks later, Fletcher Christian led his famous mutiny. Bligh and eighteen others were set out to drift in a small boat. Miraculously they drifted all the way across the Pacific. Bligh successfully sailed back to Tahiti in 1792 and then on to the West Indies, where he planted the first breadfruit tree in 1793. The tree was still standing when Queen Elizabeth II visited the Botanic Gardens in St. Vincent in 1966, for the ceremonial planting of a scion from Bligh's original transplant.[78]

The new crops were not always welcomed. Plantations expanded rapidly in South Asia during the second half of the nineteenth century. In many places—most notably Indonesia—foreign rulers had to use force to convince the natives to cultivate the crops.[79] In French Equatorial Africa (now Congo Brazzaville in part), the French outlawed hunting among the Mandja people to force them solely into cotton production.[80] Today, many of these impoverished countries depend on the crops introduced by colonialists. Central American nations that bear the scars of this history are called "banana republics," and it is no accident that these countries, growing crops not native to the Americas, have such faltering economics. By contrast, countries that benefited from this syndrome are now considered the bastions of democracy: the Netherlands, France, and England.

The plant collecting expeditions and subsequent introduction of new plants into foreign lands created the conditions for new diversity to emerge. But in the twentieth century, the nature of these expeditions and the plants sought underwent a dramatic change. About one hundred and fifty crops were circulating in world commerce. The search for and introduction of new exotic plants slowed considerably. Instead, plant explorers began to search the Vavilov centers for plant diversity—varieties of crops already known and grown—in order to introduce new genes and traits into plant breeding programs. By 1973, the need to collect and draw upon the strengths of these old varieties led Jack Harlan, by then a distinguished geneticist, to declare: "These resources stand between us and catastrophic starvation on a scale we cannot imagine. In a very real sense, the future of the human race rides on these materials."[81]

Why are these old varieties important? What could the centers of diversity discovered by Vavilov possibly have to do with the success of our agricultural system? Why value the diversity created by primitive, ancient agriculture?

Diversity. What difference does it make?

CHAPTER THREE

Value of Diversity

The fate of all forms of life . . . depends on the *continuity of variation.* At the entrance to CERES, the Controlled Environment Research Laboratory in Canberra (Australia), there is the following inscription: "Cherish the earth, for man will live by it forever." We might have said with equal justification: "Cherish variation, for without it life will perish."—Sir Otto Frankel

Little is "natural" in contemporary agriculture. It is not nature's way to allow large expanses of land to be planted to a single crop, much less to a single variety of that crop. As agriculture took hold and developed in ancient times, however, a certain fragile balance came to exist between plants and pests and diseases.

Primitive varieties—landraces—exhibited a great deal of genetic variation. A cursory look at a Neolithic wheat field would reveal differences from plant to plant. To be sure, pests and diseases struck, but their attacks were muted by the diversity and strength of the defenses accumulated by the plants during thousands of years of adaptation under agriculture. On the margin of many a field grew wild relatives of crops, which frequently crossed with those crops, infusing them with greater stamina and resistance. Moreover, fields were not contiguous; there remained ecological barriers to the spread and buildup of diseases. As a rule, diseases only rarely exploded into widespread epidemics. Crops were damaged, but not devastated.

But then the world changed—or at least our perception of it did. The event was the visitation to Ireland of *Phytophtora infestans* in the 1830s.[1]

A native of the Andes, the potato was unknown in Europe prior to the "discovery" of the New World. Potatoes were introduced into Spain in 1570 and into England and Ireland about 1590 or a few years earlier. For 250 years all potatoes grown in Europe were descendants of these two introductions.

In France, King Louis XVI became an advocate of the potato. In a neglected field near Paris he grew a wonderful crop of potatoes protected during the day by royal guards. Realizing that the peasants would be impressed by any crop so guarded, he cleverly withdrew the guards at night, allowing the peasants to raid the fields, which they did. Soon the king's goal was accomplished—all over France potatoes were growing.[2]

In Ireland the potato became the staple crop of the poor. By the 1840s the average adult was eating nine to fourteen pounds a day.[3] The summer of 1845 was a particularly good one for Irish farmers and reports from a number of counties indicated potato crops of "the most luxuriant character . . . promising abundant yield."[4]

Then it happened. On September 11, the *Freeman's Journal* announced:

> We regret to have to state that we have had communications from more than one well-informed correspondent announcing the fact of what is called 'cholera' in potatoes in Ireland, especially in the north. In one instance the party had been digging potatoes—the finest he had ever seen—from a particular field, and a particular ridge of that field up to Monday last; and on digging in the same ridge on Tuesday he found the tubers all blasted, and unfit for the use of man or beast.[5]

The potatoes in the ground as well as those already harvested began to turn black and rot. An awful stench filled the countryside. The weather was blamed. Next year would be better.

But it was not. The failure of the potato crop was a disaster for the Irish poor, who were numerous. Though three-quarters of the land was devoted to cereal crops (which were producing fine harvests), nearly all of this was exported to England or used to pay rents to landlords,[6] many of whom were foreign. At the time, four thousand people owned eighty percent of the land, while the annual earnings of a rural laborer rarely equaled the rent on a single acre of land. John Mitchel, an Irish journalist of the time, charged that "in 1847, during the Great Hunger, Ireland produced agricultural products to the value of £44,958,120, enough to feed Ireland twice over, and continued to do so, but the people starved because with this produce the rent had to be paid."[7] On occasion mobs attempted to prevent the grain from moving to the seaports for export, but such efforts were thwarted by the military.

The peasants could afford neither to keep nor to buy the grain they raised. Instead they lived on potatoes. A third of the Irish population was totally dependent on the potato for nourishment.

The British who ruled Ireland were concerned, but, as one observer noted at the time, "one must remember that the Irish had a terrible tendency to exaggerate."[8] The British did repeal import duties on grain in order to lower the price of bread. But in the best of times, the poor could not afford bread—that was why they existed on potatoes.

Next the British imported corn and stored it in warehouses. But the Irish were unaccustomed to eating corn and the mills ill-equipped to process it. Meanwhile, Irish-raised wheat was being exported to Britain in huge quantities. Irish relief societies actually began to *import* grain from England at inflated prices—grain that had been grown in Ireland and exported because the people, now starving, were too poor to pay for it!

The British opposed giving food to the starving lest it encourage the idle poor. Charles Trevelyan, the British bureaucrat in charge of the government's famine policy in Ireland, finally shut down public work programs and ceased all other government relief in 1847, declaring in effect that the famine was over and voicing his concern that "the only way to prevent people from becoming habitually dependent on government is to bring operations to a close. Uncertainty about the new crop only makes it more necessary. . . . These things must be stopped now, or you run the risk of paralysing all private enterprise . . ."[9] And with that statement, Trevelyan left Ireland to what he called "the operation of natural causes," and went on vacation to France. Knighted for his work in Ireland upon returning from France, Sir Charles Trevelyan set about writing a history of the Irish famine, which, predictably, he claimed had ended in August, 1847.

But the famine was not over. The winter of 1847–48 saw corpses lying in the streets unburied for days. By the spring of 1849, the toll had become staggering. One county with a population of five thousand had over seven hundred people die in a two-week period.[10] Some then—as now—said that the dreadful poverty of the people was caused by overpopulation. But as the population was decimated, the poverty remained. There was no appreciable rise in incomes. Considering emigration, Ireland's population for the century leading up to the famine was, according to Erna Bennett, "probably slightly higher than the British, but insignificantly so."[11] Moreover, the amount of cultivated land per capita in Ireland immediately prior to the famine was similar to that of other European countries. Ireland was more densely populated than Denmark and Prussia, but less so than England, Wales, Scotland, or Belgium.[12]

The famine continued for five years altogether. For five years Ireland's potatoes rotted. One to two million people died and as many migrated to North America, including the ancestors of two American presidents—Kennedy and Reagan—and of your two authors, Fowler and Mooney.

It was not the weather that struck down Ireland's potatoes in the 1840s; it was *Phytophtora infestans,* a potato blight. The potatoes grown in Europe—genetically limited, as we have seen—were not resistant to this disease, and their lack of resistance allowed the blight to reach epidemic proportions. Luckily not all potatoes were vulnerable. Among the thousands of distinct types in the Andes and in Mexico, resistance was located. Without it, potatoes probably would not be a major crop in the developed world today.

But the blight has consistently been blamed for the famine. As devastating as the disease was to potatoes, it was the social system, by allowing few to own and control so much, that caused the famine. How else can we explain the fact that eighty percent of the countryside was still being grazed, not cultivated, and that grain continued to be exported at a time when hundreds of thousands were perishing? Even as people were starving, Ireland produced enough food for everyone. The Irish themselves said: "God sent the blight; the English brought the famine."

Potatoes were the first crop in modern history to be devastated by lack of resistance—and the first crop to be rescued by the wealth of defenses built up over thousands of years in its center of diversity. Thus the Irish potato famine stands both as the most dramatic warning of the dangers of genetic uniformity and the clearest example of the value of preserving genetic diversity.

Differences in the performance of different crop varieties had long been noted. Various Greek and Roman writers recorded their observations on this matter.[13] But little use of this knowledge was made until the advent of modern plant breeding in the 1800s and the rediscovery of Gregor Mendel's laws of heredity at the turn of the century.

With this knowledge, plant breeders were able to use the diverse characteristics of landraces developed over thousands of years to fashion new crop varieties designed for particular situations. By carefully selecting for the desired characteristics, breeders could "weed out" unwanted traits and arrive at a "pure line," a variety that was uniform and reproduced this uniformity.

The diversity and variability of the old landraces used in early breeding programs were thus whittled down to a pure line. And often one pure line was bred with another to create a hybrid. In the field these genetically restricted varieties replaced the wide open diversity, the "harmonious disorder"[14] of the landraces.

> "A pure line mentality, convinced that variation was bad, uniformity was good, and off-types in the field somehow immoral, developed. Symptoms of the mental climate could be found in crop judging contests, ribbons awarded at county and state fairs, crop improvement associations, seed certifying agencies, and in some provisions of state and federal seed acts. It did not seem to occur to anyone that a deliberate mixture of cultivars could be a useful alternative to pure line culture. Although grain is frequently mixed in the elevator anyway, a mixture in the field was considered bad husbandry and a slightly less than mortal sin to be kept hidden on the back forty off the road."[15]

Though the new varieties were clearly superior in some respects—yield being the most obvious—they sometimes lacked the breadth of resistance, or a trait like cold tolerance, contained in the landraces. Simply put, the landraces would not have survived as long as they had under harsh conditions without fertilizers or pesticides if they had not been adapting effectively. Contributing to their success was the spatial heterogeneity provided by early farming systems. Mixed cropping made it difficult for pests and diseases to build up excessively. With the new varieties the flexibility of general adaptation and resistance was traded for something more focused and inflexible. Replacement of landraces with new, pure line varieties planted over thousands of acres opened the way for pests and diseases to attack the uniform, inbred plants. In a field of landraces a pest might gobble up one plant but find the next one different enough to be distasteful. In a field of modern varieties, if the first tasted good, they were all going to taste good.

In the 1870s coffee rust essentially wiped out the coffee industry in Ceylon (now Sri Lanka), India, East Asia, and parts of Africa. As a result, England became a nation of tea drinkers. Epidemics hit cotton in the 1890s.[16] And in 1904, an epidemic of stem rust struck the U.S. wheat crop. By 1905, what is probably the oldest program designed to develop disease resistance in this crop was begun by the U.S. Department of Agriculture.[17]

The race continued. Other epidemics followed. In 1917, "wheatless" days were declared in the U.S. in response to an epidemic. Twenty-six years later and half a world away, brown spot disease devastated the Indian rice crop, touching off the infamous Bengal famine. In the 1940s, cultivars accounting for eighty percent of the U.S. oat crop were eliminated, and oats experienced more problems in the 1950s.[18] Then in the early 1970s, corn blight struck in the U.S., sparking concern over genetic uniformity in the nation's crops. And a major failure of the Soviet wheat crop, caused in part by the large-scale planting of an inappropriate variety, precipitated the "Russian grain deal" and dramatic (and ultimately costly) shifts in American farm policy.

Each time resistance was needed. And each time it was found in the centers of diversity, in landraces that had somehow escaped homogenization, or in those crops' wild relatives.

As use of the pure line and hybrid varieties increased, so did pest and disease problems. The greater pest and disease problems grew, the more farmers turned to chemicals to solve them. In 1945, less than 200 million pounds of pesticides were employed in the U.S. Thirty years later the total had risen to 1,600 million pounds.[19]

But the chemicals did not solve the problems. It could even be argued that as pesticide use increased, so did pest problems. With all the increased "firepower" in the hands of farmers, it might be expected that the pest rebellion would have been put down. But in the last forty years the percentage of the annual crop lost to insects has doubled in the U.S. Losses due to diseases have also increased.[20] There are reasons for this.

Much of agriculture's pest control work is done by "beneficial" insects that feed on or otherwise control "harmful" insects; or by spatial heterogeneity. In a "natural" setting there are more harmful than beneficial predatory insects. If there were not, the beneficial insects would begin dying off from lack of food. A one-to-one ratio between harmful and beneficial insects would mean that the useful insect's first meal would probably be its last.

Most pesticides kill both useful insects and agricultural pests without distinction. One corporation, Rockwell International, advertised in a magazine that its pesticide would "kill every bug you've got" and all the insects pictured were creatures that never harm crops.[21] A pest problem may be alleviated temporarily by a pesticide as it kills all insects at hand. But, as Dr. Carl Huffaker of the University of California says, "When we kill a pest's natural enemies, we inherit their work."[22] When the pesticide dissipates or is washed away by rain, the harmful insects multiply rapidly. The population of beneficial insects—small to begin with and now completely decimated—cannot recover as quickly. Nor, because of their small numbers, do these beneficial insects develop resistance to chemicals as easily. The result: harmful insects return with a vengeance, this time with even less to stop them.

It is becoming common knowledge that insects are developing resistance to the pesticides that once killed them. Resistance to DDT began appearing among crop-eating insects just six years after introduction of that infamous pesticide.[23] Most people realize that today's super pest could enjoy several doses of yesterday's pesticide for dessert, ask for more and live to tell about it. Over four hundred species of pests have now developed resistance to the chemicals that once destroyed them. More than a million chemicals have already been screened for their effectiveness as pesticides and yet "the progress of resistance may outpace the discovery of effective new materials." According to Dr. Andrew Forgash of Rutgers University, "it will be increasingly difficult and expensive to design, develop, and introduce new pesticides that can stem the rise of resistant pests to the uncontrollable status, which appears to be imminent for a number of very serious pests."[24] Researchers at the University of California report a 7.7 percent increase in the number of species resistant to one or more chemicals developed between 1976 and July, 1978.[25]

Insects can also learn to evade the plant's natural resistance. In an experiment conducted by the International Rice Research Institute (IRRI, pronounced "eerie"), plant pathologists raised the troublesome brown planthopper on a rather poor quality, but brown planthopper-resistant rice variety by the name of Mudgo. The insect, which is now the most serious pest in Asia according to T.T. Chang of IRRI, was unknown to rice workers in the early 1960s. In this experiment many brown planthoppers starved to death rather than eat Mudgo. On average the first generation of planthoppers lived only 4.2 days, but that was long enough for them to produce a new generation. The new generation did not find Mudgo quite as distasteful. By the tenth generation, the planthoppers were indicating that Mudgo was all right. They lived an average of sixteen days eating nothing else, which is about as long as they lasted feasting on one of their favorite, susceptible rice varieties.[26]

With insects developing both immunity to pesticides and a taste for resistant varieties so quickly, it is little wonder that the average life span of a new cultivar has, in the memorable words of Lawrence Hills, "been reduced to that of a pop record."[27] Kenyan wheats, to cite but one example, last an average of 4.3 years before they have to be pulled off the market and replaced by a new variety.[28]

Like insects, diseases also adapt both to chemicals and to the genetic defenses of plants. Diseases mutate, developing new races to overcome the resistance of plants and the farmers' chemicals. "Race 1" of the standard wheat stem rust was identified in 1917. Fifty years later, three hundred were known.[29] Like insects, plant diseases have coevolved with their host organism. They are not in the business of becoming extinct. By adapting to their changing environment they are able to survive.[30] And survive they do.

As the pests chalk up victories on the battlefield, chemical company bravado rises in pitch. Government regulations slowing down their ability to add to the chemical load already in use, claim the pesticide manufacturers, have prevented them from winning the war. In one startling editorial in a Dow Chemical Company newsletter, Dr. C.A. Goring, director of agricultural products research for Dow said: "Given the opportunity, the web of man-made chemical technology could continue to grow in beauty and diversity creating many new and beautiful species, eliminating some old species and relegating others to more specific ecological niches." Did you catch that? "Given the opportunity," Dow's chemicals will create a few new species, eliminate some species (might we ask which?) and assign others to "more specific ecological niches." Anyone opposing this plan, according to Goring, is an "anti-technologist." After all, he continues, "the chemicals we make are no different from the ones God makes."[31]

As agriculture developed in ancient times, the balance between plants and pests and diseases rarely got too far out of line. A disease too successful would ultimately eliminate itself! Plants survived. Pests and diseases survived. With the creation of pure line varieties, however, much resistance was lost as the diversity of the landraces was reduced to create uniform varieties. Plant species that had had to rely on their own natural defenses and on mixed cropping systems for thousands of years were suddenly forced to depend on man to help them resist new or stronger pests. Thus breeding programs were established to re-insert resistance into crop varieties that were now under constant attack.

When plant breeders settle down to the long and expensive job of developing a disease resistant variety, their first line of attack is to look for that resistance in other modern varieties; then they try landraces. Modern varieties and their breeding stocks give the breeder the least trouble, because they are usually similar in other respects to the variety desired. It takes less effort to eliminate the undesirable characteristics while obtaining the needed resistance. Landraces, because they have survived so long among pests and diseases in the centers of diversity, offer a wealth of potential resistance. But they are not as similar to the end product desired and some pains must be taken to eliminate unwanted characteristics as the resistance is obtained.

Canada's famous wheats were produced by breeding varieties and landraces from Australia, England, Kenya, Egypt, India, Poland, Portugal and the Middle East.[32] And in the U.S., a Chinese spinach variety "rescued Virginia's spinach industry from ruin."[33] A thorough listing of such examples would take many books. Suffice it to say that the "primitive" varieties developed by our ancestors continue to play an integral role in the maintenance of modern crop varieties.

When all else fails and resistance or some other desired characteristic cannot be found in cultivated types, the plant breeder will turn to closely related wild or weedy plants for the needed genes. Breeders call these plants "wild relatives." There is no trace of a smile on their faces when they use the term, because working with wild relatives is often so difficult. For every desirable characteristic obtained, a number of completely unacceptable ones must be bred out. Ridding the new variety of wild and weedy characteristics can mean years of extra work for the breeder, hence the use of wild relatives in breeding programs is usually a sign of either desperation or courage on the breeder's part. Many plant breeders do not even know, could not even recognize, the wild relatives of the crops they specialize in breeding. And few are eager to find out.

The diversity of wild relatives has enabled them to survive longer than the oldest cultivated variety—and to survive without human assistance. If their genetic resistance had failed them, they would have become extinct long ago. Thus, as sources of resistance, wild relatives are a treasure.

Wild relatives have now been used in the breeding programs of virtually every cultivated crop.[34] Sugarcane is "an example of a commercial crop that has been completely salvaged" by the use of wild relatives.[35] The same could probably be said of strawberries[36] and sunflowers using genes found in North America.

Wild species from Central and South America offer the only known source of resistance to the most serious disease that strikes black pepper.[37] And in peanuts, wild species are now "the main source of resistance to pests and diseases."[38] Likewise, potato breeders are becoming more and more dependent on wild relatives. In the Federal Republic of Germany, nine out of ten potato seedlings have wild species or primitive landraces in their backgrounds.[39]

Tomatoes and tobacco simply "could not be grown commercially at all in the U.S.," according to Jack Harlan, without the resistance they have developed from wild species.[40] For at least nineteen disorders of tomatoes, wild tomato species are the principal source of resistance.[41] They have supplied resistance to leaf mold, tobacco mosaic virus, nematodes, curly top, *Septoria* leaf spot, wilt, and other diseases.[42] And they have helped extend the growing range of this crop.[43] Finally, wild species of tomato offer some interesting possibilities for future breeding work. One of the world's leading experts on tomatoes, Dr. Charles Rick of the University of California, found a number of wild tomatoes growing along the beach on one of the Galapagos Islands. Barely five meters (16.5 feet) from the high tide line, these plants were exposed to Pacific salt spray and very salty soil. Back at the University of California at Davis, Rick's colleagues found that these tomatoes could be grown hydroponically in a culture "gradually adjusted to full-strength sea water."[44]

After a four-year program which tested seventeen thousand rice accessions and over one hundred wild taxa, resistance to grassy stunt virus was found in just one population of *Oryza nivara* from India. When a new strain of the virus appeared in 1982 more screening was forced. After arduous testing, resistance was again found, again in wild species.[45] And wild species have helped give potatoes resistance to *Phytophtora infestans,* of Irish potato famine fame.[46] In 1951, an epidemic of barley yellow dwarf virus broke out in California. The search for resistance took breeders to one gene in an Ethiopian barley.[47]

Chocolate and chocolate-lovers were delivered from twin curses of witches' broom and swollen shoot, two diseases that strike cacao, from which chocolate is made. Again, wild and semi-wild species furnished resistance.[48] Wild cottons are being used in breeding programs aimed at producing varieties resistant to the dreaded bollworm and boll weevil.[49] Other wild cottons offer greater fiber strength[50] and some show promise in reducing the cause of brown lung disease which afflicts textile workers. Breeders of rubber, pineapple, cassava, and maize are seeking to introduce greater vigor into their crops by using wild species.[51] The range of soybeans, grapes, and wheat, in addition to the previously mentioned tomatoes, has been extended by employing genes from wild relatives.[52]

Wild species even show promise of helping improve the nutritional qualities of wheat,[53] rice, rye, oats, soybeans,[54] and a number of vegetable crops. And others, like some of the bramble fruits, are being dethorned through breeding programs using wild plants.[55]

Robert and Christine Prescott-Allen estimate that between 1976 and 1980, genetic material from wild relatives contributed $340 million per year in yield and disease resistance to U.S. farmers. According to the Prescott-Allens, wild germplasm has contributed $66 billion to the American economy[56]—an amount greater than the total international debt of Mexico and the Philippines combined[57]; and the comparison is not unrelated. As we shall see in following chapters, this valuable germplasm is routinely donated by Third World countries to the U.S. without any compensation or corresponding reduction of official Third World debt to the U.S.

Using wild species in breeding programs is a difficult, trying, time-consuming, and costly venture. Most breeders work under the pressure of getting new varieties out into the market—and getting them out in a hurry. Since breeders understandably shy away from using the wild relatives, it is doubly impressive that they have recently had to turn to these plants so often for such important traits in so many crops. In many instances cited above, even the primitive landraces were passed by as resources for breeding programs—rejected either because resistance could no longer be found in their ranks, or because the resistance they held was no longer effective.

Plant breeders have the unenviable job of trying to help modern agricultural crops stay one step ahead of thousands of pests and diseases. In any given year they are mostly successful. But resistance is a moving target. This year's resistant variety is next year's main course. Plant breeders have not always been able to prevent this.

In the early days of this century the old landraces provided the raw material with which plant breeders began to work. Highly variable, these primitive varieties have continued to offer much in the way of resistance and adaptability. But today the weedy and wild relatives of these crops have assumed a degree of importance that would have shocked, if not discouraged, yesterday's breeder. In truth, they have become so essential that Jack Harlan—arguably the scientist with the broadest knowledge of their role in agriculture—was moved to say that the "wild relatives stand between man and starvation . . ."[58]

Years ago, Vavilov described plant breeding as "evolution at the will of man." Like any kind of evolution, plant breeding requires variation. Artists create paintings using palettes covered with colors. Plant breeders fashion new varieties using the genetic variation within a crop. Robbing the breeder of this variation is like taking colors from the artist. It diminishes what is possible. If too much variation is lost, little or no evolution is possible. Eventually, the crop succumbs and becomes extinct.

Without the landraces and wild relatives, our modern crop varieties would be incapable of changing, of evolving, of adapting to new conditions, or stronger pests. Like so many things in this world, the new depends on the old. Without the old varieties, the new varieties could not continue. They simply could not survive. And herein lies the irony. In the long run, the future of agriculture and the very survival of crops depend not so much on the fancy hybrids we see in the fields, but on the wild species growing along the fence rows, and the primitive types tended by the world's peasant farmers in the centers of diversity. Without these wild species and old landraces, there would be no agriculture. So we turn now to what may be the most important question facing our own species: What is the state and well-being of these little known resources that stand between us and starvation?

CHAPTER FOUR

Genetic Erosion: Losing Diversity

Suddenly in the 1970s, we are discovering Mexican farmers planting hybrid corn seed from a midwestern seed firm, Tibetan farmers planting barley from a Scandinavian plant breeding station, and Turkish farmers planting wheat from the Mexican wheat program. Each of these classic areas of crop-specific genetic diversity is rapidly becoming an area of seed uniformity.

—Garrison Wilkes

John Deere is the company that makes the big green tractors seen all over the world. It is also the publisher of *The Furrow,* a colorful and lively bi-monthly magazine about farming and John Deere products. This is not a magazine to which one can subscribe. To receive it you must be placed on the list by a John Deere dealer. Presumably, most recipients are farmers.

In 1972, *The Furrow* devoted its July–August issue to the marvels of modern plant breeding. On its cover was a page from a fictitious seed company catalogue ("Created Crops, Inc.") that displayed and described a number of new crops created in the imaginative minds of *The Furrow*'s staff. There was corn-phlox, a combination of corn and the flower, phlox. "Turn your ho-hum corn fields into scenic tourist attractions. These magnificent plants glow so beautifully you'll want to sit around and watch your phlox by night." And then there were wheat-beets, a cross between wheat and sugarbeets, the newly patented plant that keeps the "heavy heads waving while the beets go on." And how about cotton-n-candy? You guessed it—cotton and sugarcane in one plant. But the winner was the Super Salad plant. Here is the description:

> Our plant engineers have finally wrapped up nature's finest salad makings into one, magnificent multi-layered vegetable plant. First an outer layer of crisp, green lettuce; next, a tender ring of onions; then, a colorful wrap of green pepper that encloses a superbly-flavored tomato. Each plant produces several bowl-sized salads, ready to eat. Takes the fuss out of salad-making. With our SUPER SALADS, you just add the dressing. Pkt. 98 cents.

And there it was on *The Furrow*'s cover—the Super Salad plant—a head of lettuce enclosing onions, peppers, and tomatoes all sliced up and ready to go, just as the Super Salad plant had grown it. It was great fun. Great fun, that is, until the orders started coming in. Over five hundred of them for seeds of the Super Salad plant alone. Five hundred orders from *farmers*.

Perhaps it should not be too surprising that so many farmers would order these seeds. Though plant breeders have not produced anything as fanciful as a Super Salad plant, amazing results have been achieved in plant breeding, particularly since the 1950s. Most famous of the wonders of modern plant breeding were the high-yielding grains, the so-called miracle wheats and rices.

It was no accident that these high-yielding varieties came into being. For this to happen, research institutes had to be established, scientists employed, distribution systems developed, and financing arranged. This network of research institutions, corporate interests, and philanthropies emerged as a response to political situations. It sought to solve problems. What were these situations? How were those problems defined? What were the goals? If we are to understand the forces that still dramatically influence agriculture, these are some of the questions that must be answered.

Between the 1940s and 1960s, international crop breeding institutes were established in Mexico and the Philippines with the objective of increasing "food supplies as quickly and directly as possible."[1] The high-yielding varieties they produced promised a green revolution in the restless and hungry countryside of the Third World. While the humanitarian goals of feeding the hungry are obvious, a major part of the impetus behind promotion of the green revolution lay in the desire to forestall revolutions of another color. China was being "lost to the communists." The British were fighting communists on the Malay Peninsula. There was trouble in the Philippines. The French were in the process of losing Indochina. The U.S.-backed government in Korea was dealing with rural uprisings.[2] And in Mexico, the Cardenas government had expropriated Standard Oil and become distinctly hostile to large landholders.

The U.S. government responded with military measures: troops in Korea, and military assistance in Indochina and the Philippines. But it was recognized that these problems stemmed from rural discontent, and rural discontent stemmed from hunger. Hunger was the ally of the communists.

Writing in the prestigious journal *Foreign Affairs,* John King echoed the prevailing wisdom of the times: "The major problem in the struggle to keep South and Southeast Asia free of Communist domination, is the standard of living of these peoples. The struggle of the East versus the West in Asia is in part a race for production, and rice is the symbol and substance of it."[3]

Closely linked with the goal of quieting political opposition in the countryside was the need to open up these rural areas to trade and development. William Myers, dean of the School of Agriculture at Cornell University and a Rockefeller Foundation trustee, returned from a trip to Mexico with John D. Rockefeller III in 1951, and observed that the Mexican economy was "handicapped by hundreds of thousands of uneconomic farm units . . ." In a letter to the president of the Rockefeller Foundation that year, Myers reported that "these small farms cannot make use of improved agronomic practices because they have no surplus above family needs to sell to finance such improvements."[4]

The Rockefeller Foundation, long active in Third World rural health, educational, and agricultural programs, was encouraged by Myers to play a key role in developing the agriculture-oriented response to the political crises of the 1950s and 1960s. The foundation had no interest in organizing peasants or in land reform—the topics on the minds of the rural poor. It had accepted the conclusion of its Survey Commission on Mexican

Agriculture that the "most rapid progress can be made by starting at the top and expanding downward."[5]

The project in Mexico included strengthening the National School of Agriculture at Chapingo and working closely with the government from an Office of Special Studies within the Mexican Ministry of Agriculture. It culminated with the establishment of the International Maize and Wheat Improvement Center (CIMMYT).

The approach was similar in Asia. At the top of the priority list was rebuilding the College of Agriculture at the University of the Philippines. The foundation gave Cornell University a major grant for establishing a Southeast Asia program to develop American experts and train students from Southeast Asia. That same year (1952), the U.S. Mutual Security Agency offered a contract to Cornell to develop the agriculture school at the University of the Philippines. Cornell had already been invited to establish agricultural colleges in Thailand. Between 1951 and 1960, over six hundred Thais (including a hundred extension workers) were given agricultural training courtesy of the U.S. government.[6]

Building from this base, the Rockefeller Foundation and Ford Foundation proceeded to establish the International Rice Research Institute (IRRI) in the Philippines.

Thus the stage was set. The infrastructure needed to produce a major transformation of Third World agriculture was in place. This transformation promised not only to increase food production and still the radical fervor in Asia and Latin America, but also to bring those regions into the fold as participating members of the market economy.

In a relatively short time, research institutes had been established, scientists employed, agricultural leadership developed, and teachers and extension workers trained. Lest we be misinterpreted—we are not describing a conspiracy. We are simply recounting some of the major steps taken in response to the political crisis arising in the Third World. No one—certainly not the plant breeders and extension workers, and probably not even the government planners or Rockefeller Foundation officials—could have anticipated even half of the effects the green revolution would have in the countryside.

Scientists at CIMMYT, IRRI, and other institutes set out to increase the productivity of Third World agriculture by breeding and distributing high-yielding crop varieties, principally cereal grains. CIMMYT and IRRI plant breeders utilized genetic material containing dwarfing genes, which encouraged the shifting of biomass from stems and leaves into grain.

The technology developed during World War II to produce bombs made possible the production of nitrogen fertilizers after the war. As it turned out, the new varieties were fertilizer-responsive: they were capable of taking the added nutrients in fertilizers and translating them into greater yields. Use of the new seeds and fertilizers resulted in yield increases ranging from ten to a hundred percent or more—no small feat. "Rice politics" hit the campaign trail, and in short order the new varieties were being grown on millions of hectares.[7]

While food production increased on green revolution lands, hunger persisted. In fact, a series of International Labour Organization studies concluded that hunger and malnutrition increased most rapidly precisely in these areas.[8] To some, like Dr. Keith Griffin who authored a major report on the green revolution for the UN Research Institute for Social Development, it appeared that the new seeds were not "neutral."[9] From the beginning they did not grow as well for the poor farmers as for the richer ones. Achieving high yields required fertilizer and irrigation. Fertilizer and irrigation nourished weeds as well as crops, creating the need for herbicides. And pests found the uniformity of new varieties appetizing, which necessitated the use of insecticides as well. Farmers lacking access to capital to buy these items were simply left in the dust. In rice-growing regions, economies of scale were not as readily possible as in wheat-growing regions, where some farmers cultivated sufficiently large areas to take advantage of their relative wealth and reap the benefits of the new technologies. Would-be consumers found that food produced by fancy, imported technology was too expensive to buy when their annual income was only a couple of hundred dollars. These were not the results intended by the well-meaning and dedicated plant scientists. The results were simply the inevitable consequence of the types of technology developed and promoted, and the world into which the technology was introduced.

Nevertheless, governments, research institutes such as CIMMYT and IRRI, agricultural extension workers, aid agencies, chemical and farm machinery corporations, and humanitarian groups alike all pushed the new technology, though sometimes for very different reasons. Seeds were given away. Loans were made for fertilizers and equipment. In the Philippines, government loans through the "Masagana 99" program in 1981 were given only to farmers who agreed to plant a government-recommended variety. Only ten varieties (all produced by IRRI) were on the program's list for the whole of the Philippines,[10] a country which stretches a distance equivalent to that from Rome to southern Sweden.

In Iran, large landowners who mechanized their farms were even exempted from a land reform act.[11] But while the new wheats from Mexico were successful in the low plains along the Persian Gulf and the Caspian Sea, "their cultivation led to a complete fiasco in the uplands."[12] By 1978, even IRRI publications were acknowledging that because the new technology had been employed irregularly over the countryside, there had been a "widening of income differentials . . . among communities and regions."[13] And within communities there were reports that the rich were getting richer and the poor poorer due to differences in their access to the whole package of new technologies.

Dr. Hermann Kuckuck of the Technological University of Hannover observed that "in countries where the increase in production was particularly striking, such as in Pakistan and India, the process of polarization in the social sector was intensified and led to new social tensions."[14] As a World Bank report explained: "Large scale farmers generally acquire knowledge of such technologies more quickly, and because they have better access to the working capital needed to utilize these technologies more fully, they capture the early and largest gains from innovation. . . . At least in the short run, relative distribution of income worsens as between large-scale and small-holder farmers."[15] Like other reports, this one from the World Bank held out the hope that as time passed all would adopt the new seeds, thus lessening the current disparities. But presently this remains no more than a hope, and perhaps an unrealistic one at that.

The green revolution answered the problem of hunger and rural unrest with increased production, not with land reform or employment projects; essentially it offered a technological solution to a social and political problem.

The green revolution increased yields, at least in some areas of some countries. Some countries reduced or eliminated food imports. A few have been able to export a little food. But growing more food only solves the problem of hunger if the hungry eat that food. Just because a country raises production and ceases importing food does not mean that the hungry have been fed. It may mean that they have no money to buy food. It can also mean that the added grain production is going to feed animals—where about half of the world's wheat, corn, barley, oats, rye, and sorghum goes—animals that eventually produce meat for those wealthy enough to consume this kind of protein.[16]

The relationship between the new seeds and the availability of fertilizers in the early days of the green revolution was somcthing akin to the relationship of the chicken and the egg. The fertilizers made the new varieties possible. The new varieties made the fertilizers necessary.

As fertilizers and irrigation became more common, distinctions between different environments were evened out. The extreme example of this today is the greenhouse, where virtually all conditions are controlled. In such circumstances, adaptability on the part of the plant loses much of its importance.[17] Uniformity becomes more possible, more attractive.

By creating one basic environment, the combined forces of cultivation, fertilizers, herbicides, insecticides, and irrigation all helped set the stage for the introduction of uniform varieties "suitable" for vast expanses of land. The economic incentives and social prestige to be gained from using these modern techniques addressed the human element.

The spread of the new varieties was more dramatic than anything that had ever happened in agriculture before. Within a decade new varieties of wheat and rice were being grown on nearly fifty-five million hectares (135 million acres) in the Third World.[18] By 1976, forty-four percent of all land in wheat and twenty-seven percent of the land in rice in developing countries was planted with the new, miracle varieties.[19] Today they can be found growing in virtually every corner of the world, by the largest, richest farmers, and by the most destitute peasant farmers—all trying to coax a more bountiful harvest from their plot of earth, all willing to trade in their old lettuce, onion, tomato, and pepper varieties for the new Super Salad plants.

"Modern" plant breeding efforts begun in Europe and North America in the nineteenth century initiated the process of replacing the traditional landraces with new, inbred varieties. By the early twentieth century many of the European landraces—types grown there for hundreds and even thousands of years—had disappeared. Initially, no one paid much attention to this process. The old varieties were, after all, being replaced by newer and *better* varieties; this was progress. No one seemed to realize that after countless generations, the traditional varieties were domesticated. Unlike weeds or wild species, they had become dependent on people for their existence. Replacement was and is simply another word for extinction. Landraces that disappeared were gone forever.

Europe and North America were rich and important sources of crop genetic diversity, but few of today's major crops originated there. The real wealth of genetic resources was in Vavilov's centers of diversity in Third World countries, protected there by the economic "backwardness" of those countries, and by transportation systems that prevented the marketing of new seeds being developed in the U.S. and Europe. But even the centers of diversity were not impregnable.

H.V. Harlan and M.L. Martini sounded the first alarm:

> In the great laboratory of Asia, Europe, and Africa, unguided barley breeding has been going on for thousands of years. Types without number have arisen over an enormous area. The better ones have survived. Many of the surviving types are old. Spikes [of barley] from Egyptian ruins can often be matched with those still growing in the basins along the Nile. The Egypt of the Pyramids, however, is probably recent in the history of barley . . . In the hinterlands of Asia there were probably barley fields when man was young. The progenies of these fields with all their surviving varieties constitute the world's priceless reservoir of germ plasm. It has waited through long centuries. Unfortunately, from the breeder's standpoint, it is now being imperiled . . . Today, in years of shortage, the French supply their dependent populations with seed from California. Arab farmers . . . import seed from Palestine. In a similar way changes are slowly taking place in more remote places. When new barleys replace those grown by the farmers of Ethiopia or Tibet, the world will have lost something irreplaceable. When that day comes our collections, constituting as they do but a small fraction of the world's barleys, will assume an importance now hard to visualize.[20]

That day has now come, perhaps in part because so few paid much attention to these prophetic words in the U.S. Department of Agriculture's 1936 *Yearbook of Agriculture;* and in part because the green revolution accelerated the process of genetic erosion more than even Harlan and Martini could have anticipated.

How much has been lost? It is impossible to answer that question, since we do not know how many varieties existed. Information regarding traditional varieties that once grew in centers of diversity is extremely scarce, which means that comparisons with how many remain today may not be very illuminating.

Fortunately, the United States Department of Agriculture assembled detailed listings of varieties being sold by commercial U.S. seed houses in the early twentieth century. These lists have made possible the first comprehensive study of the degree of losses in one country. This study, undertaken by the Rural Advancement Fund International (RAFI), compared early USDA lists with varieties now held by the U.S. National Seed Storage Laboratory, the major U.S. seed bank and perhaps the largest such facility in the world. In the case of apples and pears, the old lists were compared to holdings in both U.S. and European facilities.[21]

Let us stress that the study and its results are flawed in several ways. While the early USDA lists were well prepared (and attempted to account for synonyms—one variety being known by several names), they dealt only with varieties being sold commercially, except in the case of fruits. In the early 1900s, many, many varieties had been developed and were being saved by individual farmers and gardeners outside the sphere of commercial seed companies, which were not nearly as prominent then as they are today. Thus many varieties were omitted from the original USDA list. In other words, this study indicates how much of what seed companies were offering some eighty years ago has been lost. It does not indicate the scope of losses among varieties not sold by seed companies back then. Based on our knowledge of what the U.S. now has stored in seed banks, we would conclude that losses among these noncommercial varieties have been at least as great.

When we speak of "loss," we are ultimately talking about extinction. We have presumed that any variety found on the early USDA lists that could not be found stored in U.S. seed banks today is probably extinct. We suspect a few are not—a few might be found in European seed banks and a few more varieties have surely been saved year after year by generations of a family of dedicated farmers or gardeners. (One of the authors, for example, "discovered" three apple trees of a variety previously believed extinct, the Bonum, in a small commercial orchard operated by the Willie Jones family in the hills of Virginia. And it is now known that the Seed Savers Exchange based in Iowa has quite a few varieties.) But such cases are rare. And if such varieties were ever needed, plant breeders might not be able to find them, or might not even know they existed. In effect, they are "functionally extinct."

However, there is a silver lining to this darkest of clouds. Varieties are unique combinations of genes. It is possible that some, most, or all of the genes of an extinct variety still exist in another variety, though not in that particular combination. Thus, strictly speaking, the table below is an indicator of the loss of varieties, not genes. Given the fact that none of the "functionally extinct" varieties was studied prior to disappearance, however, it is impossible to say there have been no losses of genes. And given the magnitude of variety loss, one might even argue that many distinct genes and characteristics have been lost.

Surveying some seventy-five types of vegetables, RAFI found that approximately 97 percent of the varieties given on the old USDA lists are now extinct. Only 3 percent have survived the last eighty years. Table 1 gives the results of the study by vegetable.

The RAFI study also covered extinction of apple and pear varieties. Of 7,098 apple varieties in use between 1804 and 1904 in the United States, 6,121 or 86.2 percent have been lost. Of 2,683 pear varieties in use during those same years, 2,354 or 87.7 percent are now extinct. Included among these is the Ansault pear, of which one of America's premier fruit experts, U.P. Hedrick, once said: "In particular, the flesh is notable, and is described by the term *buttery,* so common in pear parlance, rather better than that of any other pear. The rich sweet flavor, and distinct but delicate perfume contribute to make the fruits of highest quality . . . Ansault should find a place in every collection of pears for home use."[22]

These losses of fruit and vegetable varieties are staggering. And perhaps most frightening of all is the fact that while most of the crop diversity that once existed in the U.S. is now gone, the U.S. has never had the crop diversity that many other countries in centers of diversity have had. The U.S. lost a lot, but it had less to lose. If this study of the U.S. is any indication, extinction rates in other countries during the last century may be even more awesome. Yet day by day, the losses climb. More and more varieties become extinct, never to be seen again.

CROP BY CROP LOSSES

In the summer of 1982, Roland von Bothmer of the Swedish University of Agricultural Sciences and Niels Jacobsen of the Royal Veterinary and Agricultural University in Copenhagen set forth to explore North America for wild *barley.* They carried with them the handwritten notes of a former teacher of theirs, who had seen it growing in two locations in and near Santa Barbara, California, in the 1950s and 1960s. The first, they discovered, "is now the site of the First Presbyterian Church" and its parking lot. No barley there. The second place had been an unkept cemetery. Von Bothmer and Jacobsen pressed forward. Upon finding the cemetery they discovered that it had become a "highly-cultivated park

Table 1
U.S. Vegetable Varieties Lost, 1903–83

Vegetable		Total 1903 varieties	1903 varieties in U.S. NSSL collection	Varieties lost (%)
Artichoke	*Cynara scolymus*	34	2	94.1
Asparagus	*Asparagus officinalis*	46	1	97.8
Beans				
Runner bean	*Phaseolus coccineus*	14	1	92.9
Lima bean	*Phaseolus lunatus*	96	8	91.7
Garden bean	*Phaseolus vulgaris*	578	32	94.5
Beets	*Beta vulgaris*	288	17	94.1
Mangel beet	*Beta vulgaris*	178	3	98.3
Broccoli	*Brassica oleracea* var. *botrytis*	34	0	100.0
Brussels sprouts	*Brassica oleracea* var. *gemmifera*	35	4	88.6
Burnet	*Sanguisorba minor*	1	0	100.0
Cabbage	*Brassica oleracea* var. *capitata*	544	28	94.9
Cardoon	*Cynara cardunculus*	6	1	83.3
Carrot	*Daucus carota*	287	21	92.7
Cauliflower	*Brassica oleracea* var. *botrytis*	158	9	94.3
Celeriac	*Apium graveolens* var. *rapaceum*	25	3	88.0
Celery	*Apium graveolens* var. *dulce*	164	3	98.2
Chervil	*Anthriscus cerefolium* *Chaerophyllum bulbosum*	8	0	100.0

Table I
Continued

Vegetable		Total 1903 varieties	1903 varieties in U.S. NSSL collection	Varieties lost (%)
Chicory	*Cichorium intybus*	17	3	82.4
Chives	*Allium schoenoprasum*	1	1	0.0
Chufas	*Cyperus esculentus*	2	0	100.0
Collards	*Brassica oleracea* var. *acephala*	28	5	82.1
Corn				
Field corn	*Zea mays*	434	40	90.8
Popcorn	*Zea mays*	48	0	100.0
Sweet corn	*Zea mays*	307	12	96.1
Corn salad	*Valerianella olitoria*	21	1	95.2
Cress	*Lepidium sativum*	39	2	94.9
Cucumber	*Cucumis sativus*	285	16	94.4
Pickling cucumber (gherkin)	*Cucumis anguria*	10	2	80.0
Dandelion	*Taraxacum officinale*	25	0	100.0
Eggplant	*Solanum melongea*	97	9	90.7
Endive	*Cichorium endiva*	64	4	93.7
Horseradish	*Amoracia rusticana*	1	0	100.0
Kale	*Brassica oleracea* var. *acephala*	124	9	92.7
Kohlrabi	*Brassica oleracea* var. *gongiyodes*	55	3	94.5
Leek	*Allium ampeloprasum*	39	5	87.2

Table I
Continued

Vegetable		Total 1903 varieties	1903 varieties in U.S. NSSL collection	Varieties lost (%)
Lettuce	*Lactuca sativa*	497	36	92.8
Martynia	*Proboscidea louisianica**	4	0	100.0
Muskmelon	*Cucumis melo*	338	27	92.0
Mustard	*Brassica juncea*	44	5	88.6
Okra	*Abelmoschus esculentus*	38	4	89.5
Onion	*Allium cepa*	357	21	94.1
Orach	*Atriplex hortensis*	5	1	80.0
Parsley	*Petroselinum crispum*	82	12	85.4
Parsnip	*Pastinaca sativa*	75	5	93.3
Pea	*Pisum sativum*	408	25	93.9
Peanut	*Arachis hypogaea*	31	2	93.5
Peppers	*Capsicum annuum*	126	13	89.7
Radish	*Raphanus sativus*	463	27	94.2
Rampion	*Campanula rapunculus*	1	0	100.0
Rhubarb	*Rheum rhaponticum*	35	1	97.1
Roquette	*Eruca sativa*	1	0	100.0
Rutabaga	*Brassica napus* var. *napobrassica*	168	5	97.0
Salsify	*Tragopogon porrifolius*	29	2	93.1
Scolymus (Spanish salsify)	*Scolymus hispanicus*	1	0	100.0
Skirret	*Sium sisarum*	1	0	100.0
Sorrel	*Rumex acetosa*	10	0	100.0

Table 1
Continued

Vegetable		Total 1903 varieties	1903 varieties in U.S. NSSL collection	Varieties lost (%)
Spinach	*Spinacia oleracea*	109	7	93.6
New Zealand or Malabar spinach	*Tetragonia expansa*	1	1	0.0
Squash	*Cucurbita* spp.	341	40	88.3
Sunflower	*Helianthus annuus*	14	1	92.9
Swiss chard	*Beta vulgaris*	23	1	95.7
Tomato	*Lycopersicon esculentum*	408	79	80.6
Husk tomato	*Physalis philadelphica*	17	2	88.2
Turnip	*Brassica rapa*	237	24	89.9
Watercress	*Nasturtium officinale*	2	2	0.0
Watermelon (and citron)	*Citrullus lanatus*	223	20	91.0

*Martynia indicates accessions not varieties

with richly-cut lawns," where wild barley has "no possibility of growing."[23]

In the global scheme of things, North American barley germplasm is not terribly important (though prehistorically there was a North American barley domesticate). More diversity is located in Africa and the Near East, where churches and cemeteries in no way constitute the most serious threat.

With the establishment of crop breeding institutes turning out modern varieties within the actual centers of diversity, the genetic erosion noticed by Harlan and Martini in the 1930s became an unending avalanche—an avalanche destroying the priceless and innumerable genetic resources found in these centers.

By the 1970s, indigenous *wheat* varieties in Greece had virtually disappeared except in remote mountain areas.[24] It had been required by law in Greece to grow CIMMYT wheat varieties. In Thessaly and Macedonia, fewer than ten percent of the varieties grown are indigenous.[25] India's native wheats began disappearing in the 1940s and 1950s. When wheat varieties from CIMMYT in Mexico arrived in the 1960s, the process was "virtually completed."[26]

In North Africa, CIMMYT wheats as well as varieties from Kenyan breeding programs are replacing native varieties. Traditional varieties are most likely to be found in the Atlas Mountains and around the oases of the Sahara, if at all.[27] There are reports that even the wheats and barleys of the oases are virtually gone.[28] These same ubiquitous new Kenyan varieties were found by Dr. Hermann Kuckuck of the Federal Republic of Germany over a decade ago "in very remote areas of Ethiopia, accessible only by mules."[29] Meanwhile, the native wheats of the Nile Valley will soon be gone, replaced by modern varieties provided by a government program.[30]

Recent expeditions to the Himalayas have discovered that modern wheat varieties are even finding their way into isolated valleys long protected by the mountains. In Nepal, the green revolution wheat, Sonalika, is a common sight. The new varieties are now grown on eighty percent of the land, and have displaced about that proportion of the traditional wheats.[31]

Surveys of Sicily's wheats in the latter part of the last century turned up fifty native varieties. More recent studies indicated that fewer than forty percent have survived.[32]

The remnants of Turkey's once rich diversity of wheat are now to be found only in "remote villages in the Pontiac and Taurus mountains," according to a sixteen-year-old survey.[33] Once thought to be safe, Afghanistan's wheats are now disappearing. This country's immense diversity, noted in *The Agriculture of Afghanistan* by N.I. Vavilov and D.D. Bukinich in 1926, is falling before imported CIMMYT varieties.[34] According to Erna Bennett, who recently retired as crop genetic resources officer with the United Nations, "Thousands of tons of seed have been imported. Introduced varieties now predominate in many parts of the country."[35] During the past several years, war and the resultant mass migration of a sizeable portion of the rural Afghan population into Pakistan have produced renewed concern over one of the world's richest reservoirs of wheat diversity. Our inquiries—made to both the Afghan and Russian governments—about the safety of this diversity and about any collecting expeditions that had been mounted in light of the obvious threat—have gone unanswered. We fear that much has been lost, perhaps nearly all.

In Spain, Pakistan, Eastern Europe, Libya, and Iran, the story is the same. "By the 1970s," according to the International Association of Plant Breeders (ASSINSEL), almost half the total wheat area of the developing countries was under varieties developed under CIMMYT guidance."[36]

How many varieties of wheat existed in the first place? What percentage has been lost? No one has clear answers. Scientists and plant breeders who collected wheats thirty or forty years ago can safely make statements about the severity of the losses since then. But the number of wheat varieties that existed before the process of replacement began simply could not be counted. No one ever tried. (For the story of some recent ill-fated attempts, see chapter eight.)

How serious are the losses? Again, who can say? We do not know the characteristics of varieties now extinct. But we remember the sobering story of a wheat collected in 1948 by Jack Harlan in Turkey. Arriving in the U.S., it was given the plant introduction number 178383. No name was deemed necessary. Harlan described it thusly:

> It is a miserable-looking wheat, tall, thin-stemmed, lodges badly, is susceptible to leaf rust, lacks winter hardiness . . . and has poor baking qualities. Understandably, no one paid any attention to it for some 15 years. Suddenly, stripe rust became serious in the northwestern states and P.I. 178383 turned out to be resistant to four races of stripe rust, 35 races of common bunt, ten races of dwarf bunt and to have good tolerance to flag smut and snow mould.[37]

Harlan's miserable wheat is now used in all breeding programs in the northwestern states of the U.S. and saves farmers millions of dollars each year. Can we safely lose thousands of varieties of wheat today with the assurance that we will not need them in the future? It is not likely.

The farmers in the Pacific Northwest are lucky. They had a close call. Barley growers know what it feels like. There are lots of kinds of barley. Over 6,500 different kinds would be more than enough, you would think. But when scientists at the U.S. Department of Agriculture screened this many looking for resistance to barley yellow dwarf virus, they could locate only one gene that could confer resistance.[38]

Rice breeders have faced similar situations in the past, although the

IRRI has done a commendable job of collecting endangered rice varieties in recent years. Norman Myers notes that:

> A few years ago, the famous "miracle strain" of rice in the Philippines, IR-8, was hit by tungro disease. Rice growers switched to a further form, IR-20, whereupon this hybrid soon proved fatally vulnerable to grassy stunt virus and brown hopper insects. So farmers moved on to IR-26, a super-hybrid that turned out to be exceptionally resistant to almost all Philippines diseases and insect pests. But it proved too fragile for the islands' strong winds, whereupon plant breeders decided to try an original Taiwan strain that had shown unusual capacity to stand up to winds—only to find that it had been all but eliminated by Taiwan farmers as they planted virtually all their ricelands with IR-8.[39]

The replacement and extinction of traditional rice varieties has not always been a pretty process. Formosa is a case in point. Soon after the Japanese acquired the island they undertook a ten-year survey of the native varieties, and found 1,997. By 1910, a "Native Rice Improvement Program" was under way. "Poor quality" rices were eliminated and in their place farmers were encouraged to grow selected varieties that appealed to the Japanese consumer. Red rice, a hardy, drought resistant type, was popular on the island but considered of poor quality by the Japanese. Thus, as Andrew Pearce tells the story, "a harsh campaign against red rice and other varieties was conducted by the police. By 1920, the number of native varieties had been reduced to 390."[40]

During the past fifty years, Indian farmers may have grown over thirty thousand different varieties of rice. With the spread of new varieties, H.K. Jain of the Indian Agricultural Research Institute now estimates that in fifteen years just ten varieties "may cover as much as 75% of the total rice acreage in the country."[41]

In Senegal, Gambia, Guinea-Bissau, Guinea, Sierra Leone, Liberia, Ivory Coast, Burkina Faso, Ghana, Mali, Togo, Benin, Nigeria, Niger, the Cameroons, and Chad, *African rice,* which was domesticated by Africans independently of Asian rice, is disappearing. In the late 1960s, international aid agencies arrived in Upper Volta. They dammed a river and constructed irrigation systems to support the new Asian rice varieties they introduced. Yields of the new rice outstripped those of the African rice tenfold. Farmers hardly knew what to do. But in 1971, everything changed. African diseases struck the Asian rice. Yields dropped to familiar levels. But with all the money invested in irrigation, seeds, fertilizer, and equipment, who could afford the old yields?[42] Unfortunately, only ten percent of the rice to be found by then was African rice.[43] It was too late to turn back.

The spread of high-yielding *millet* hybrids has been slowed considerably due to their susceptibility to diseases.[44] It was thought that the traditional varieties were under "no risk of erosion" as Sir Otto Frankel put it.[45] But several years after that assessment, millet was given a high priority listing for collection by the International Board for Plant Genetic Resources (IBPGR) in Rome. The reason: immediate threat of genetic erosion.

With *sorghum*, change came overnight with the introduction of the hybrids in 1956. The hybrids increased yields by fifty percent.[46] Four years later eighty percent of the area planted to sorghum in the U.S. was devoted to the new hybrids.[47] These hybrids were marketed in Africa where Jack Harlan tells us the impact on traditional varieties has been "considerable."[48]

In chapter two, we highlighted the marvelous diversity of *potatoes* and their equally varied names. Dr. Carlos Ochoa of the International Potato Center in Peru is one of the world's leading authorities on the tuber. He is a breeder, an explorer and a collector of traditional varieties. Potatoes that are "pig droppings" to some are little treasures to Ochoa. Ochoa reports that genetic erosion (extinction) is taking place "very actively" in parts of the Andean region of South America, and he describes the situation in Mexico and Guatemala as "very alarming." Like others, Ochoa is finding that new varieties are replacing traditional potatoes even in extremely remote localities.

Dr. Ochoa is a sensitive scientist caught in a dilemma to which many of his colleagues elsewhere are blind. He feels a need to breed new, improved potato varieties. But he fully understands the necessity of preserving the traditional varieties. In Ochoa's case the dilemma can become quite personal, as is evident in the following excerpt from a letter we recently received from him.

> I remember that near to 25 years ago I was exploring the Northern of Peru. At that time, still was possible to find dozens of interesting primitive potato cultivars, which 20 years later was more difficult to find such variability, many of them, like 'Naranja' for instance, probably are extinct. The main reason, I am sorry to say, is the introduc-

> tion of "Renacimiento" [rebirth], one of the varieties that I breed, long time ago, for this country.[49]

In northern Peru, Ochoa found that in two areas where he had previously collected forty-five different varieties, none could be found by the early 1970s.[50] Rambramina, Ila, Clavelina, Chilopa, and Montanera—these are the names of potato varieties that can be added to the list of those that will never be seen again, replaced and driven extinct by Renacimiento.[51]

On the island of Chiloe off the coast of Chile, diversity in potatoes was noted by Charles Darwin on a stop he made there. In the 1920s, a Russian botanist found some two hundred types. Twenty years later the number had been cut in half. Subsequent surveys have found even fewer.[52]

Wheat, barley, rice, millet, sorghum, potatoes—millions of people's lives depend on these staple crops. All are experiencing serious losses of the diversity they acquired over thousands of years. But "genetic wipe-out," as this process has been termed, is not limited to these staples, mostly cereals. Vegetable and fruit varieties are being reduced as well.

Diversity in *okra* is found in Africa and India. In the Sahel of Africa, dried okra is sold in the markets. In Brazil, India, and West Africa, it is eaten fresh. And in the southern U.S., it is eaten fresh, pickled, fried, and boiled. In Louisiana, it is a main ingredient in "gumbos." Traditional varieties are being replaced by modern varieties like Emerald, Louisiana, and Goldcoast. Clemson Spineless is a popular okra variety developed in South Carolina and now sold by two dozen American seed companies. Its name notwithstanding, Clemson's okra is as aggressive as its athletic teams, and is now eliminating native varieties in Africa.[53]

Old varieties of *sugar peas,* perhaps similar to those grown across Europe at the beginning of the century, were completely wiped out in the Pienine Mountains of southern Poland when new varieties were introduced by government agricultural offices in 1976.[54]

In Turkey, a country of immense genetic wealth in many crops, native *sugar beets* are being replaced by imports of new models with names like Detroit Globe.[55] Sugar beets are not alone. Variation in *lentils* is being threatened by a government program.[56] The chickpea is yet another crop once thought safe but recently designated as a priority for collection by the International Board for Plant Genetic Resources.[57] Foreign *beans* are threatening Turkey's "remarkable genetic variation" with "rapid extinction," according to Frankel's survey published in 1973.[58] Although the country was once a center of diversity for flax, great expanses have now been planted to a single variety. Fruit tree nurseries established by the government in Turkey thirty years ago now account for eighty percent of what is grown there. Lists of recommended government varieties have been prepared and promoted with the resulting extinction of many traditional *peach, pear, sweet cherry,* and *apricot* varieties.[59] With the completion of new road systems, even remote areas need no longer be self-sufficient in fruit. Now that fruit (of the government's varieties) can be trucked in, many local growers have stopped raising fruit, which is eliminating yet more local varieties.[60]

In Europe the *cabbage family,* which as we have seen includes broccoli, cauliflower, and Brussels sprouts, is under severe attack from modern varieties. In cauliflower, the losses have already proved costly.

Cornwall, in the extreme southwest of England, has sported a flourishing *cauliflower* industry for years—ever since the railway system opened up the rest of the country to Cornwall's winter and early spring produce. Cornwall was the home of Old Cornish cauliflower, a variety brought to Cornwall from Italy about 150 years ago. A number of different types evolved in Cornwall and evidence suggests that they were resistant to ringspot, a disease common to the region.[61]

In the 1940s the British Ministry of Agriculture and Sutton's, a British seed company, obtained and began to breed French cauliflower types, which also originated in Italy. ("This in itself is a fascinating story, but may well contain libellous components," says Dr. Peter Crisp of the National Vegetable Research Station at Wellesbourne.) Much of the breeding work was done at Seale Hayne, an agricultural college in Devon. The varieties developed by Sutton's and the ministry were introduced, and by the 1950s had completely replaced Old Cornish cauliflowers. Soon thereafter, Cornish farmers must have begun to notice ringspot disease, and today it can be found on virtually every cauliflower growing in the region during winter. The new varieties simply were not resistant enough to prevent the spread of this disease, which is now so common as to discourage the planting of Brussels sprouts and winter cabbage in Cornwall as well.[62]

The new varieties caused the extinction of the best (and perhaps only) real source of resistance to the disease—the Old Cornish cauliflowers.[63] We will never know what other valuable traits may have disappeared with Old Cornish cauliflowers. The center of diversity in cauliflower is the eastern Mediterranean, though much diversity developed throughout Europe. But diversity in this center is disappearing as the Italian cauliflower industry grows[64] and as Japanese varieties are introduced.[65]

As their name implies, *Brussels sprouts* originated in Belgium. In the words of the International Board for Plant Genetic Resources, new hybrids are causing "massive genetic erosion in this crop."[66] Replacement of early varieties of Brussels sprouts is "virtually complete in Japan, Korea, North and South America, and Northern Europe."[67]

N.L. Innes, in a study of Brussels sprouts varieties in the United Kingdom, found that while the number of varieties offered by commercial seed companies had grown from fifty in 1965 to seventy-one in 1974, the increase was something of an illusion. Personal inquiries made by Innes to breeders revealed that the increase in varieties was due entirely to introduction of new *hybrids*. In fact only eighteen of the varieties offered in 1965 were still available in 1974. Many of the new types had been derived from a few inbreds. Attempts to obtain old varieties listed in the catalogues over the last twenty years failed. Innes was forced to conclude that many of the traditional varieties in the United Kingdom "have already been lost," and also found that twenty-four non-hybrid cabbages had been eliminated from the catalogues between 1965 and 1974.[68]

Writing from the Institute for Horticultural Plant Breeding in Wageningen, Holland, Dr. Q.P. van der Meer says: "A nearly complete substitution of open pollinated varieties by hybrids took place around 1975 for Brussels sprouts and around 1980 for *white cabbage*. The same process is going on with respect to *onions* and *carrots*. . . . The new varieties are provided by private Dutch seed firms, e.g., Bejo, Sluis en Groot [Sandoz], Royal Sluis, and Broersen."[69]

As difficult as this might seem to believe, many companies rid themselves of the seed stocks of varieties they have discontinued. And we know of no government that systematically collects commercial varieties as they are deleted from the catalogues. Thus, a variety offered by your favorite seed company might well become extinct one or two years after that company drops the variety from its catalogue.

This litany of the replacement and extinction of traditional crop varieties at the hands of modern ones produced by crop institutes, government agencies, and the seed industry could continue long into the night. In addition to the crops already cited, scientists have documented this process with *amaranth, apple, lima bean, bitter gourd, Chinese cabbage, cotton, cucumber, eggplant, flax, forage grasses, maize, mung bean, oats, onion, peas, pumpkin, soybean, spelt, squash, sweet pepper, sweet potato, tomato, watermelon, yam,* and a number of other crops.

In reviewing much of the scientific literature on this subject written during the past twenty years, one finds replacement of traditional varieties as the cause of extinction over ten times more often than any other reason. While other factors may cause extinction of old varieties, clearly *the major force causing the loss of our agricultural heritage is the introduction of new varieties* produced by professional breeders. This in turn leads to the destruction of mixed cropping systems in which landraces are often found and in which they are most appropriate. To reiterate, plant breeders do not produce the new varieties in order to drive older varieties into extinction. But in the absence of effective programs to collect and preserve the old varieties when new types are introduced, extinction is the inevitable result.

Most commentators examining the loss of genetic resources explain that the cause is "modernization of agriculture," a term that sounds as antiseptic as it does inevitable. This explanation gives no insight into the planning and efforts that certain interests made to create the preconditions for "modernization" and no inkling of the forces that persuaded and cajoled Third World farmers to get "modernized."

The new varieties were and are at most only a tool—one element of a broader phenomenon whereby subsistence agriculture is being challenged, and its practitioners integrated into the market economy. What has this meant at the local level? Certainly the new seeds called for different skills from the Third World farmer, who suddenly had to learn how to use pesticides and fertilizers, to manage irrigation systems, and to master the machinery required. The larger harvests and the necessity of selling them required new methods of crop storage and new skills in marketing and finance. For many a Third World farmer, it was not just a new way of farming that had to be learned, but a whole new world view that had to be assimilated.

Commercializing Third World agriculture put pressure on local customs. Communal ownership of land and the notion prevalent in Central America that seeds should always be given as gifts, and never sold, were just two examples of traditions that came under attack.

More subtle at first was the fact that the new inputs—seeds, pesticides and fertilizers, and even the expertise—originated in the industrial sector. Formerly self-provisioning economies became dependent on outsiders very quickly as "the location of decision-making of farming practices (tended) to shift from farmers to non-farmers."[70] And as this transformation took place it became easier and easier for government planners to listen to scientists and corporate interests rather than to farmers, for it was no longer the farmers who were in control of agriculture.

The seeds came with the genetic code of the society that produced them. They produced not just crops, but replicas of the agricultural systems that produced them. They came as a package deal and part of the package was a major change in traditional cultures, values, and power relationships both within villages and between them and the outside world. It is important to note that this process sometimes ripped apart and destroyed local cultures. In other cases it was decay of a traditional culture, and fascination with things western, that allowed the new varieties and the culture they brought with them to gain entrance.

So it was not something as clean as the "modernization of agriculture" or as simple as the importation of new varieties that was behind this tremendous surge in genetic erosion since the 1950s, but an exceedingly complex process involving high stakes politics, fancy new technologies, and desperately poor peasant farmers.

The foregoing description of this process is based largely on statements of those in the industrialized world most intimately involved in it. There should not be anything very controversial about it. But to some the very mention of this process is enough to cause hypertension. Talk about it at all and you are likely to have all sorts of charges hurled in your direction, the most common being: "You just want to *go back*" (with the implication that this would mean starvation for millions). "Just how many people are you going to condemn to starvation?" Harold Loden, then executive vice-president of the American Seed Trade Association, demanded to know in a televised debate with us. Analyzing the process which threatens genetic diversity no more implies wanting to "go back" than caring about the preservation of the *Mona Lisa* means wanting to go back to the sixteenth century. Quite the reverse. Analysis of threats to diversity is a key prelude to confronting the future.

We have thus far concentrated on the production and distribution of new seed varieties as a cause of extinction in old varieties.

Of the remaining causes of extinction, two stand out as important. The first is familiar, being a variation of what we have already been examining. In many centers of diversity the introduction of modern "cash crop" varieties—of wheat, for example, or virtually any other crop—is often reason enough for a peasant farmer to abandon cultivation of a traditional variety of an entirely different species—perhaps turnips—in order to achieve the higher yields and income associated with the initial adoption of new varieties. *Amaranth, barley, cocoa, fodder beets, carrots, potato, spelt, forage grasses,* various *tree fruits, rye, wheat,* and particularly *maize* are crops that can be shown to have lost traditional varieties due to their replacement by modern varieties of other crops.

This problem is particularly serious with *legumes.* Legumes are rich in protein, and in many Third World countries their proteins are needed to complement those found in grains. Millions of people who depend on the root crop cassava rely on legumes for their principal source of protein. Nevertheless, green revolution varieties of other crops have frequently replaced local legume crops, not only causing their extinction, but also depriving local people of a crucial source of protein.[71]

Habitat destruction is the other leading cause of genetic erosion. We have not yet stopped inventing new ways to destroy the habitats of plants that nourish us. Habitat destruction is the chief enemy of wild species, which, as we have seen, are increasingly important in plant breeding.

The building of Egypt's Aswan Dam and subsequent flooding of an enormous area by Lake Nasser put some cereal strains out of business for good. And when the people of the Wadi Halfa area were relocated, overgrazing by their animals in their new home reduced and perhaps eliminated the wild sorghum that had grown there.[72]

Overgrazing is taking its toll on wild potatoes, as are logging and the expansion of human settlements in the Andes.[73] Norman Myers, author of *The Sinking Ark,* found that in Turkey "wild progenitors of several cereals find sanctuary from grazing animals only in graveyards and castle ruins."[74] Overgrazing by domestic goats also poses the greatest threat to wild tomato species in coastal Peru and, according to Professor Charles Rick of the University of California, "it is becoming more serious in the highlands, as well, with the worsening economic condition in those areas."[75] Rick cited the impact of industry and land clearing for agriculture or housing as additional causes of tomato extinction.[76]

People in many parts of Africa and Asia still use wild plants for food and derive important nutritional value from them. Nutritionally important wild plants in Indonesia include Gnetum, a leafy vegetable originally from Assam, India; *Sauropus;* and the South Pacific *Hibiscus manihot.* But in Indonesia, as in so many other parts of the world, the allure of all things "modern" is evident, as wild lands harboring such plants are converted to fashionable imported vegetables like cauliflower, carrots, and cucumber.[77]

The relentless advance of deserts—an area twice the size of Belgium becomes desert every year—poses the real threat to an African tree called Laperrine's olive. Tuareg tribesmen are cutting the tree for wood, as desertification reduces their wood supply. It is thought that the olive might prove of value in olive breeding programs if it survives.[78]

Cocoa, from which chocolate is made, is severely threatened in the wild. Traditional Criollo types of fine quality and exceptional flavor were once found throughout Central America and much of South America. Many were replaced by other high-profit crops—bananas in Costa Rica and Honduras, cotton in Nicaragua. Others were replaced by "improved" varieties.[79] Deforestation due to oil exploration by Texaco, Phillips Petroleum, and others in the late 1960s and early 1970s destroyed additional populations,[80] almost certainly without the knowledge of these corporations. Few oil or timber companies survey the areas they are about to disturb or make efforts to preserve natural vegetation.

The process of cocoa extinction continues today, though there is little left to destroy. Native varieties have been wiped out in Central America.[81] Only one extensive, pure planting of a traditional cocoa variety is known to Dr. Jorge Soria, a cocoa expert and assistant deputy director of Costa Rica's Inter-American Institute for Cooperation on Agriculture. That planting of a type called Porcelana is found in the southern part of the Lake Maracaibo region in Venezuela.[82]

And if lack of appreciation for the fine qualities of Criollo cocoas were not enough, overexploitation threatens another tree. When we were growing up, "Chiclets" was a popular brand of chewing gum. Chiclets and other chewing gums are made from chicle, a gum tapped from the sapodilla tree in Mexico and Guatemala. The demand is apparently high and younger trees are now being tapped without being allowed to rest and recuperate between tappings. Many are dying as a result of this practice,[83] presenting, as any child would tell you, a sticky situation for the chewing gum industry.

Dams, occasional famines that cause people to eat up their seeds, church parking lots, and oil drilling that affects certain trees, all constitute a certain degree of danger to genetic resources. But the single greatest threat to our agricultural heritage comes from agriculture itself, from the replacement of traditional seeds and farming practices by modern, inbred crop varieties. These new varieties pose a clear and present danger to the diversity that agriculture has produced during its long history. And with the advent of biotechnology and its promise of even higher yields (and other as yet undreamed of "improvements"), the threat to traditional varieties would seem to be even more severe and more immediate.

Which crops are most threatened? Primarily the crops where active breeding programs are producing new varieties. And, of course, these are the crops most important to human survival.

COLLECTING PRIORITIES

In 1974, the Consultative Group on International Agricultural Research, a collection of crop breeding institutes including CIMMYT and IRRI, established an organization "to promote an international network of genetic resources centres to further the collection, conservation, documentation, evaluation, and use of plant germplasm . . ."[84] The new body, the International Board for Plant Genetic Resources (IBPGR) is headquartered in the buildings of the Food and Agriculture Organization of the United Nations, across a broad avenue from the Circus Maximus in Rome. (The organizations are more thoroughly examined in chapters seven and eight.)

Relying heavily on the recommendations of advisory committees and various experts, IBPGR has formulated a list of crop-collecting priorities to guide conservation activity. In general, the factors considered are the level of risk to the crop and its wild relatives; the crop's economic and social importance; the materials that need collecting; the needs of plant breeders; and the quality of present collections. The list is occasionally updated, and is made both by crop and by region, but there is nothing scientific or sacred about IBPGR's list. It is based on educated guesses. And they are probably the best guesses that can be made, even if sometimes reflecting a bias toward industrialized countries' needs.

More than fifty crops have been assigned IBPGR's highest priority rating at the time of writing, in one or more regions:

amaranth
apple
avocado
barley
bitter gourd
black gram
breadfruit and jackfruit
broad bean
cabbage family
cassava
chayote
chickpea
citrus fruits
cocoa

coconut
coffee
cowpea
dessert banana
eggplant
finger millet
foxtail millet
greengram (mung bean)
groundnut
kangkong
lentil
maize
mango
muskmelon, canteloupe
oilseed brassicas
okra
onion
peach palm
pear and quince
pearl millet
pepper
Phaseolus (common) beans
quinoa
radish
rice
sorghum
soybean
spinach
squash family
starchy banana and plantain
sugar beet
sugarcane
sweet potato
taro and aroids
tomato
wheat
yam
yardlong bean

This frightening list of crops for which genetic diversity and thus future viability is threatened should be read over slowly. It should be studied. One should ponder what it means for these crops to have ended up on this list.

Lest the length of this list dull one into thinking that IBPGR has simply dumped every agricultural crop it could find—including a few you have never heard of—into its high priority list, below are those crops designated by IBPGR as being of the second highest priority in at least one region:

almond
apricot
bambara groundnut
bottle gourd
cashew
cherry
cotton
date
fig
grain amaranth
moth bean
oil palm
papaya
peach and nectarine
potato
rice bean
rubber
walnut
winged bean

The third group contains a number of important crops like oats, rye, pea, sunflower, olive, safflower, jute, pineapple, plum, strawberry, grape, and others. There is a fourth priority group, and then there are various crops not listed at all and under study, including many tree fruits and nuts, forage crops, and plants used for medicinal purposes.

A handful of crops—wheat, *Phaseolus* (common) beans, cassava, sweet potato, coffee and tomato—are singled out as being of the highest priority globally.

In recent years the loss of crop genetic resources has prompted sometimes frantic searches for needed germplasm in the Third World centers of diversity. With the people of the world from Boston to Bangkok dependent on thirty kinds of plants for the bulk of their sustenance, one does not have to be a nature-lover to be concerned about the extinction of traditional crop varieties and their wild relatives.

UNIFORMITY—THE PLAGUE OF SAMENESS

The danger faced by all of these crops—that genes necessary for their future evolution and survival might be lost—is compounded by two additional problems: the common use of one gene for resistance in breeding programs, and the high level of genetic uniformity in farmers' fields.

In the end, all plant breeding programs live by the bottom line. Plant breeders are under continual pressure to turn out new varieties for the marketplace. Thus, they take shortcuts unknown to our ancestors.

Typically, today's plant breeder will search for one major gene to confer resistance for the new variety. Frequently, resistance in a traditional landrace is not nearly so simple. Resistance may be the product of a complex of genes, literally hundreds of genes working together. Breeding in this kind of resistance is too time-consuming, complex, and costly to the modern breeder. But it is effective. And the resistance produced is long lasting.

By utilizing one-gene resistance (even if unintentional or if made necessary by the pressure to get something on the market quickly), the plant breeder gives the pest or disease an easy target. It has only to overcome or find a way around that one line of defense. As Erna Bennett points out, this type of breeding "is not without practical disadvantages, of which the most generally admitted is the regular occurrence of resistance breaking pathogen strains. As a result, breeding resolves into a step by step evasion of the pathogen . . . There is no ultimate vision of permanent or stabilized resistance . . ."[85]

The use of one gene for resistance, one gene which is routinely overcome by pest or disease, results in that gene being "used up."[86] It no longer provides resistance. The pathogens have its number. For years there have been suggestions that plant breeders not release varieties having such one-gene resistance. Speaking of tomatoes, Pelham in 1966 asserted that this practice was resulting in the "gradual frittering away of the few forms of resistance available." This policy, he declared, was an "unforgivable waste of the world's natural resources."[87]

In the process of going after the single gene for resistance, the gene-complex—the whole set of genes that can provide stable resistance in a landrace—is often ignored, and sometimes destroyed, despite its representing "all the plant breeding work carried out by Nature over thousands of years."[88]

The loss of crop genetic resources through extinction and the squandering of the remaining resources through one-gene resistance breeding reduce the odds of our being able to counter pests and diseases successfully in the future. But not only are we eliminating genetic material that might be needed to confer resistance if an epidemic strikes; we are also increasing the chances of it striking by our use of a very narrow genetic base in our crops. The number of varieties grown in the fields of the world's modern farmers represents but a drop in the ocean of crop plant diversity. The fact that the potatoes of Europe traced their ancestry back to two *phytophthora*-susceptible introductions meant that the genetic base of potatoes was narrow and open to attack. Though corn had been considered a low-risk crop,[89] the fact that all the hybrid corn in the U.S. in 1970—several hundred varieties—had the same type of cytoplasm (used to reduce the cost of hybrid seed production) made that entire crop vulnerable to any disease that could exploit that uniformity. The corn plants were millions of sitting ducks.

A certain level of risk exists when a farmer plants a pure line variety in the field instead of a mixture of varieties. Some components of a mixture might be able to resist a pest or disease attack. But with a uniform crop in the field, if one goes, they all go.

That risk grows when the farmer's neighbor plants the same variety. But when virtually every farmer plants the same variety or group of varieties,

Table 2
Genetic Uniformity of Major United States Crops
Extent to Which Small Numbers of Varieties Dominate Crop Average

	Major Varieties	Acreage (%)
Bean, dry	2	60
Bean, snap	3	76
Cotton	3	53
Corn*	6	71
Millet	3	100
Peanut	9	95
Pea	2	96
Potato	4	72
Rice	4	65
Soybean	6	56
Sugar beet	2	42
Sweet potato	1	69
Wheat	9	50

*Corn includes seeds, forage, and silage.

the risk becomes dangerous indeed. Two years after the 1970 corn blight, the U.S. National Academy of Sciences published a 307-page study with the simple title, *Genetic Vulnerability of Major Crops*. The study revealed that the U.S. was shockingly dependent on a handful of varieties of its major crops (Table 2). The study concluded that U.S. agriculture was "*impressively uniform genetically and impressively vulnerable*" [emphasis added].[90]

Meanwhile, in the aftermath of the devastating blight, the hybrid corn seed industry was busy boasting that it had taken only one year to correct the problem and get resistant varieties back on the market. But larger, unanswered questions loomed. Industry scientists had known that the hybrids were vulnerable ever since reports of the susceptibility of hybrid corn to this disease had been published nine years earlier, but had not corrected the problem.[91] If there had been no warning that the epidemic would occur in 1970, could we expect any warning the next time? Was the lesson to be learned from the uniformity of the corn crop and the resulting corn blight that the industry can respond quickly, or that a blight can respond quickly? Were not other crops similarly uniform and similarly vulnerable? Were not U.S. varieties replacing diversity with uniformity all around the globe? And finally, was there any assurance that the seed industry would be able to find a quick solution the next time? In fact, is there such a thing as a solution to a narrow genetic base for a crop?

To be sure, these are questions that plague breeders of other crops as well. Seventy-five percent of the world's supply of soybeans are now produced in the U.S. Until recently, virtually all of this crop could trace its ancestry back to six plants from the same area of China. Like corn before the blight, all sorghum in the U.S. has one type of cytoplasm and can be traced back to one plant.[92] The same could be said of the cytoplasmic uniformity in Europe's sugar beets.[93] Much of the crested wheat grass, a forage species, in the U.S. can trace its parentage back to a single introduction.[94] All of the grass *Digitaria decumbens* in Central America is probably descended from a single parent plant cloned in 1940.[95]

Although the National Academy of Sciences study showed less overwhelming uniformity in wheat, the situation is still alarming. Consider the wheat variety Kalyansona. About the time of the NAS study, "different selections from the original cross from which this variety is derived . . ." occupied some "60 percent of the area in the world grown under spring wheat."[96] Since then, diversity in wheat used in the U.S. has probably increased, but fewer landraces are being used in the Third World.

The problem extends to ornamental trees—all three hundred thousand Bradford pears are propagations from the same tree noticed by the director of the U.S. National Arboretum in 1950.[97] And all of Mr. Burpee's hybrid zinnias are descendants of one tiny flower found in row 66 of Burpee's fields at Santa Paula, California, in 1948.[98] Even your morning cup of coffee can probably be traced back to one tree in the Amsterdam Botanical Garden from which sprang the South American coffee industry.[99]

And the situation in Europe is as distressing as that already described. Piecing together comments made by the many European scientists we have talked to over the past decade, we can reasonably say that the degree of uniformity in Europe's major crops rivals that of the U.S.

While there is obviously a great deal of uniformity in terms of the varieties grown, there are also similarities between these varieties. Many of the predominant varieties share certain traits conferred by the same gene or genes—stringlessness in beans and some dwarfism in rice and wheat for example.[100] Thus the different varieties are not all that different in important respects.

Growing such uniform cultivars over vast areas is potentially dangerous, notes D.J. van der Have.[101] It is precisely this type of uniformity that permitted the ravages of the corn blight.

The danger comes from modern varieties that have been "purified," in Sir Otto Frankel's words, varieties with a "minimum of genetic variation."[102] In some crops, like sugarcane, potatoes, and many fruits, the sexual mode of reproduction has even been suppressed. Cuttings or grafts are made, clones are produced and the crop reproduces itself virtually without the opportunity of breeding and the introduction of new genes as we know it.[103]

"The restriction of the genetic base of production itself," claims Heslop-Harrison, "can be seen in the recommended variety lists of the advanced countries, where in response to special demands—such as those of the food freezing industry or determined by the packaging requirements of supermarketing—fewer and fewer genotypes [different types] contribute more and more to total production."[104] Such lists narrow down dozens or even hundreds or thousands of possible candidates to a mere handful. The North Carolina Extension Service handout for gardeners recommends only three varieties of garden peas, and just one variety of cauliflower. These lists are not what they used to be. Over the years they have become shorter and shorter, as if the government and agricultural establishment were searching for the one perfect variety.

A U.S. Department of Agriculture list of recommended fruits published in 1897 included more than 275 different varieties of apples. This is probably more apples than are now available from nurseries. It is nearly twice as many as offered by Southmeadow, the Michigan nursery that boasts the largest selection of apples in the U.S. Today, by contrast, recommended variety lists published by the country's major apple producing states basically contain the same two dozen or so apple varieties.[105]

European Economic Community governments have gone one step further with the publishing of a "Common Catalogue."[106] Varieties not listed therein are deemed inferior and cannot be sold legally by seed companies. In practice these are the traditional, non-patented varieties which offer competition to the patented varieties owned and sold almost exclusively by big corporations. The continued existence of these varieties will depend on quick work and perpetual cultivation by preservation societies and gardeners. Most other people are unaware of it when one of their favorite varieties becomes de-listed, and are thus ill-prepared to save seeds they do not have. Many varieties—indeed up to three-quarters of all those presently grown in Europe, according to Erna Bennett—will become extinct within ten years!

FORFEITING THE FUTURE

Fifteen thousand years ago our ancestors were using thousands of kinds of plants for nourishment. The transition to agriculture necessarily caused that number to decrease. As people began to domesticate crops, the number of plant sources they depended on decreased. But at the same time something marvelous began to happen: the diversity within each crop increased as the domesticated crops came to fit into the myriad environments to which they were exposed.

Today we are entering a third major phase in our own relationship with plants. When people were hunters and gatherers, life was sustained by many plants, none truly domesticated. Early farmers reduced the number of species used, but traditional farming practices steadily increased the diversity within those species. *Today we are rapidly narrowing and destroying that diversity.*

The process of extinction of agricultural crop varieties began in the nineteenth century, picked up speed in the early part of the twentieth, and has spread like a raging cancer during the past thirty years. Enough of the diversity created by plant evolution and thousands of years of agriculture has already been destroyed for Erna Bennett to suggest that we should refer to Vavilov's centers as "centers of former diversity."[107]

There can be no doubt that the major cause of this genetic wipeout is the replacement of traditional varieties by cultivars mainly developed by crop breeding institutes and large multinational seed companies. These institutes and companies also distribute the seeds, sometimes with the help of governments and aid agencies.

Distribution of the new seeds would not necessarily cause extinction of the traditional varieties, were those varieties collected as they were replaced. But the crop breeding institutes have only been collecting the old varieties seriously since 1970. And seed companies have not yet acknowledged that their responsibility may include seeing that the seeds they displace not become extinct.

Recently seed companies have collaborated to lobby for seed patenting laws—laws allowing a company to obtain a "patent" (or patent-like certificate) on a new variety of lettuce or tomato, for instance. To be enforceable, these laws require that the new, patented variety be internally consistent—uniform. This kind of uniformity and the ongoing quest for greater and greater uniformity pleases both lawyers and pests, and is yet another factor contributing to the narrowing of the genetic base of our crops.

It is more important, however, to understand that patenting laws are being promoted by corporations and sympathetic governments alike as a tool to help increase seed exports. As a U.S. Department of Agriculture statement put it, seed patenting laws "facilitate international trade."[108]

We cannot be sure that these laws do indeed foster increased exports of seeds. But for the moment we are willing to accept industry and government claims. Seed exports from the U.S. have doubled since passage of its law in 1970. And it does make sense that armed with patent protection, a seed company would be more enthusiastic about opening up new markets than it might be if it faced competitors marketing the same varieties. No competition is the best of all worlds, and with patents a seed company can have just that. Advertising pays off. Sales and profits jump.

So far, so good. But wait. If the replacement of traditional crop strains by new varieties is causing many of those "heirloom" varieties to become extinct, won't increases in seed exports quicken that process? If patent laws encourage increased seed exports, do they not (in the absence of effective conservation measures) also cause more extinctions? The seed industry says no, but is disinclined to elaborate. We see no alternative to concluding that seed patenting laws are contributing to the extinction of plant genetic resources.

To prove this point to the satisfaction of those whose business it is to sell seeds would require only that they yield to simple logic. Yes, they acknowledge that the introduction of new varieties is causing extinctions. And, yes, they claim that patenting encourages the export of those new varieties—as they see it, that is one of patenting's chief virtues. In acknowledging these two points but denying the connection between patenting and extinction, people in the seed industry are simply whistling a merry tune as they pass the graveyard at night.

Early in 1983, the whistling rose an octave or two when Harold Loden, the head of the American Seed Trade Association, made the fantastic claim to a U.S. television audience (Phil Donahue's interview program, "The Last Word") that there was no problem with the loss of genetic resources—everything needed had already been collected and stored![109] (See chapters eight, nine and ten for an analysis of the state of collections and the effects of seed "preservation.")

In the twentieth century we have come to influence virtually all of evolution on this planet to some extent or another. In cultivated plants, evolution takes place in the fields of plant breeders and in their private stocks. If, as officials of the seed industry imply, we know what we are doing, then all will be well.

But are we ready for the staggering responsibility of guiding future evolution on earth? The first rule of successful tinkering is to save all the pieces; yet in agriculture we are discarding the pieces before we even know their value or their role. We are burning books that we have not yet read. We are like the English monarch who directed that straight trees be destroyed but that trees with curved boughs be encouraged, because they were better for constructing hulls for sailing ships his country would need. We, too, must prepare for the future—but we must bear in mind that we do not know what the future will bring.

We do not presume to know, for example, how the human race will deal with the ultimate demise of the petrochemical age. We know we will have to, because the supply of nonrenewable resources is finite. Our generation may not have to face that problem, but some future generation will.

Sensing that the dislocations and adjustments will be tremendous and painful, some simply choose to deny their inevitability. Try telling a plant breeder that we will not always have oil-based fertilizers and pesticides to use on our heavily dependent (even addicted) crops, and, likely as not, the response will be, "Yes we will. We *have* to. How else can we feed the world's population?"

We do not know the answer to that question. But we do insist that, no matter how pressing the human need for an inexhaustible supply of oil-based agricultural inputs might be, simple logic and the events of the last decade tell us that we cannot depend on them forever. In the long history of agriculture, chemical-dependent farming may well come to be seen as a passing fad.

While praying for some as yet undreamed of solution to the problem of producing great quantities of food without great quantities of nutrients, we might do well at least to save those crop resources adapted to limited input agriculture—the varieties that have evolved and survived for thousands of years without our post-World War II chemicals.

These traditional varieties may not be a total solution. But in a world with less and less oil, modern varieties addicted to oil will not be a solution either. To suggest otherwise is a cruel hoax.

Future generations will deal with these problems either with or without the crop genetic resources that exist today. We assert it is our moral and evolutionary responsibility to see that future generations have these resources to use or not, as they deem appropriate.

But whether or not we salvage and save enough diversity to allow future generations to mold crops to suit their needs should be of secondary concern. There is an even more fundamental question: Will we save enough diversity to ensure the very survival of our major food crops?

The genetic diversity being lost today is the foundation of future plant breeding, of future plant evolution. If enough diversity is lost, the ability of crops to adapt and evolve will have been destroyed. We will not have to wait for the last wheat plant to shrivel up and die before wheat can be considered extinct. It will become extinct when it loses the ability to evolve, and when neither its genetic defenses nor our chemicals are able to protect it. And this day might come quietly even as millions of acres of wheat blanket the earth.

"Who would survive if wheat, rice or maize were to be destroyed? To suggest such a possibility would have seemed absurd a few years ago. It is not absurd now," warns Jack Harlan. "How real are the dangers? What is the potential magnitude of the disaster? One might as well ask how serious is atomic warfare? The consequences of failure of one of our major food plants are beyond imagination."[110]

CHAPTER FIVE

Tropical Forests

In the woods we return to reason and faith.

—Emerson

The British biologist J.B.S. Haldane was once asked what he could tell about the nature of the Creator from examining His creation. He is reputed to have replied that the Creator must have had "an inordinate fondness for beetles."[1] Haldane knew what he was talking about: there are around a million species of beetles. And many, if not most, live in the world's forests, along with the majority of other plant and animal species. The Creator evidently liked forests as well.

Were you to set out in search of the world's largest beetle, you would find it in a tropical forest. Forests lying in the tropics cover nearly eight million square miles, an area slightly larger than the South American continent. The forests of the tropics range from dry scrubland to wet evergreen jungle, but about sixty percent of their total area is in "closed" broad-leaved forest and grassland/tree formations. Together these can be considered tropical moist forests, and in this chapter, we shall emphasize these, since they harbor tremendous animal and plant diversity.

No one knows how many species of plants and animals exist in the world. Somewhere between three and ten million would be a good guess. Probably two-thirds of all the species on earth live only in the tropics—most in the tropical forests.[2] The British Isles boast some 1,450 species of trees, shrubs and herbs—only 350 more than have been recorded on less than a square mile in Colombia.[3] Colombia may have twice as many plant species as Ecuador.[4,5] Ecuador has as many plant species as all of Central America.[6] Central America is the size of France, but the tiny country of Panama contains as many plant species as all of Europe.[7,8]

According to the naturalist Dr. Norman Myers, the tiny state of Brunei in northwestern Borneo is believed to contain two thousand tree species, whereas Holland, seven times as large, possesses only thirty.[9] "A small forested volcano, Mount Makiliang in the Philippines," says Myers, "contains more woody plant species than the whole of the United States."[10]

The tropical forest teems with more life than any other place on earth. The luxuriant foliage of Amazonia provides shelter for one in five of the world's bird species.[11] A single tree may be home to two thousand species of insects.[12] The waters of the Amazon river system are home to nearly as many varieties of fish as are found in the Atlantic Ocean.[13]

Such exuberant diversity is indicative of a very complex ecosystem—and there is not just one tropical moist forest ecosystem, but many different and distinct ecosystems.

A great deal of precision and specialization is often required for tropical ecosystems to function. One plant considered unimportant to humanity may provide food to a bird or insect that serves as the sole pollinator of a valuable species of fruit. The huge *Casearia* tree of Costa Rica, for example, provides food to twenty-two species of birds during an annual time of fruit scarcity. Without this tree, some bird species would disappear, and plants dependent on these birds for pollination or seed dispersal would also vanish.[14]

The acacia tree offers a fascinating glimpse into the often hidden ways in which plants and insects depend on each other in the forest habitat. One type of acacia lacks effective means of defending itself against pests. Were it not for some very specialized ants that live on the trees, the acacia would be an easy target. The ants hollow out thorns on the tree and raise their young inside. They feed on the acacia's budlike leaflet tips. No harm done. In return they ferociously attack any other insect that dares set foot on the tree. Vines and other plants that touch the acacia are immediately sawn off by the ants, allowing the tree to grow unrestricted.[15] These acacias and ants have coevolved strong dependencies on each other; in effect this acacia tree has been "domesticated" by the ants.[16]

Many tropical seeds simply will not germinate unless they have passed through the digestive tract of certain animals.[17] Some plant species are still holding on for dear life after having coevolved in such a way with mammals that became extinct during the Pleistocene, thousands of years ago.[18] The Brazil nut depends for reproduction on rodents which chew its seeds and break down the hard outer covering.[19] The survival of such plants depends on the preservation of the habitat of their animal assistants.

Recently scientists have speculated that the large number of hollowed-out trunks found in tropical trees might even serve an important function. The trunks provide homes for animals and, in so doing, allow the trees to benefit from the nitrogen and minerals from the nests and from defecation inside the hollow trunks.[20]

In its undisturbed state, ninety percent of the forest is in the "mature stage." When a large old tree falls, taking others with it, gaps are created. Here the struggle for survival becomes most intense as plants put their energy into growth, diverting it away from producing toxins for protection. Not surprisingly, browsing animals are known to prefer the plants that grow in these gaps.[21]

A few species can be found widely dispersed throughout a large tropical forest. But the great majority of species is found only in the isolated pockets to which they are well-adapted. One-hectare plots separated by just a few kilometers may exhibit a fifty percent difference in tree populations.[22] Furthermore, the number of individuals of a given species may be quite limited, frequently only one or two per hectare (2.47 acres).[23] In other words, the diversity of the tropical forest is widely dispersed and much of it is localized and thus rare. In some situations this adds extra precariousness to an ecosystem already extremely fragile. Not all tropical forest ecosystems are that delicate, however. Complexity can also spell strength and resilience. It certainly is not the case that any disturbance of a tropical moist forest means destruction or degradation.

Still, of all this tropical diversity, we know very little. Only half a million of the several million species in the tropics have even been named by scientists. There are just fifteen hundred systematic biologist-taxonomists in the world qualified to classify the remaining species.[24] Were they all to drop everything, move to the tropics, and successfully find and classify a new species each day, the task would take at least four and a half years—with no time off for good behavior. Even once they had accomplished this, we still would not know what value the species might have for us, or how the newly-named plants fitted into the environment. We would simply have a list of names.

Nevertheless, the story could end here without much harm were it not for the fact that tropical forest lands are shrinking. Guesses as to how fast range widely, as do conclusions about how long it will be before tropical forests are gone altogether. It is not uncommon to hear fifty years or less given as the life expectancy for tropical forests.

The varying estimates and confusion have been the result of poor and incomplete data. The best information available today indicates that doomsday for tropical forests will not come as soon as we had once thought. The "closed" forests are shrinking at the rate of 0.6 percent a year. The "open" forests are disappearing at a slightly slower rate, according to the UN Food and Agriculture Organization.[25] But even this estimate can be misleading, for the losses are not equally dispersed.

Some forests have experienced few losses. Some—protected by terrain and distance from human populations—could not easily be destroyed even if a huge effort were mounted to do so. But others face real threats and may soon be gone. Furthermore, the figures quoted above refer to "deforestation," which the FAO defines essentially as "clearing." Logging of selected trees, which can (but does not always) seriously damage the forest and cause the loss of genetic resources, is not included in this estimate. Thus, the real losses are significant—and much greater than an average forest shrinkage figure would indicate. Given the fact that some of these forests have evolved continuously for sixty million years, current deforestation rates in some countries are ample cause for immediate concern.[26]

Deforestation is not new. It preceded multinational timber companies by centuries. Understanding how and why tropical forests are destroyed today requires us to take a brief history.

THE PLANTATION LEGACY

Long before timber companies entered tropical forests, the trees were being cleared for plantations. Columbus introduced sugarcane to the New World. Sugarcane had made its way from its home in Southeast Asia to the Cape Verde Islands off the coast of Africa.[27] Sugar was considered so valuable in Europe that it was often part of the dowries of queens. In northeast Brazil it found conditions to its liking. By the end of the sixteenth century, Brazil had 120 sugar mills and was the world's largest producer. Huge areas of forest were cleared. And perhaps even larger areas were cut to provide fuel for the mills. Laboring in the fields were thousands of slaves brought from Africa.

Sugar spread. By 1666, Barbados had eight hundred sugar plantations and eighty thousand slaves. Sugar invaded the Caribbean islands one by one, destroying forest lands and encouraging the introduction of more slaves. In 1786, Haiti imported twenty-seven thousand slaves, and the following year forty thousand were brought in. Slave revolts four years later in Haiti prompted a sugar boom in Cuba.

Prior to the introduction of sugar, it was reportedly possible to walk the length of Cuba and never leave the shade of trees.[28] By 1900, half the forests were gone. Sugar, ranching, and ship building for the Spanish navy had taken their toll. Cuba was forced to become a large buyer of U.S. timber.

In Brazil, the natives' bouncing balls made from a tree gum caught the attention of Europeans and soon brought big changes to Amazonia. The Europeans took the gum back home and discovered that it would rub out lines on paper. They called it "rubber."[29] In 1839, Charles Goodyear developed a method of processing rubber which made it stronger and less sticky.[30] Uses for the new material abounded and natives were dispatched to search for and tap the one or two rubber trees that grew per hectare.

A rubber "boom" followed, but few benefited. Local Indian populations were decimated. Half a million peasants—descendants of slaves brought in to grow sugar—died on their way to the Manaus area of the Amazon or while out rubber-tapping there. The region experienced no lasting economic development. In 1876, Henry Wickham with the help of Tapuyo Indians collected seeds along the Tapajos River and transported the delicate seeds out of the Amazon, across the Atlantic to Kew Gardens in England.[31] Seedlings were started and two thousand were shipped to Sri Lanka (then Ceylon), Malaya, and Java. According to records, no more than twenty-two seedlings made it to Malaya.[32] It was simply a matter of luck that Wickham had not brought with him from the Amazon a virulent disease that could wipe out rubber plantations. Wickham had collected his seeds in just one small area, near Boim at the confluence of the Tapajos and Amazon rivers. Nevertheless, the twenty-two seedlings that arrived in Asia were sufficient to begin the rubber industry there,[33] despite their obvious lack of genetic diversity.

When rubber plantations were established in Asia, the Amazon's boom collapsed, and its share of the world market plummeted from one hundred to two percent.[34] The Amazon once again slipped into obscurity.

Cacao, another boom and bust crop for Brazil, experienced a similar fate. More forests were cleared. This time Africa got into the act of growing an imported crop, and by 1920, Ghana was the top producer.

Coffee was next. By then slavery was on the way out and rich landowners in Brazil and elsewhere found it cheaper to pay subsistence wages than contend with slaves. Millions of people became dependent on the vagaries of the international coffee market. Even love was affected: in Colombia, the marriage curve was observed to coincide closely with the coffee price curve.[35]

Bananas were introduced to the Caribbean, where they remain an important export crop today. A similar phenomenon was taking place in the Philippines and Indonesia. Sugar, abaca, coffee, indigo, and tobacco were raised for export. And in every case, forests were cleared.

But there is a more important lesson than merely that large areas of tropical forest were cleared before our time. The introduction of export crops like sugar brought with it a new economic and political system. Lands were seized, plantations were established, and enormous numbers of people were enslaved to work on the plantations.

When the institution of slavery eventually ended, power relations between rich and poor changed but little. For many, the only employment available was on plantations. Those who set out in search of land had to eke out a living on poor soil or eroding slopes—the plantations having been established on the best lands.

Export agriculture in the Third World has always demanded cheap labor, and lots of it. Lacking access to capital or land, a poor family's only asset was labor. Lacking any systems to provide for the aged, a couple's only security was children. And so the poor multiplied.

From slavery and plantation days to the present there have been few opportunities for the multitudes impoverished by this history to acquire enough land and other resources to escape poverty. Land ownership patterns remain heavily skewed in many Third World countries, with a handful of families often controlling most of the land while the vast majority own little or no land at all.

This is the present situation—a clear result of colonialism. And it is this that now poses perhaps the greatest threat to tropical forests.

In country after country, thousands of people left landless and jobless in the aftermath of the plantation economy are invading the tropical forest. To be sure, people have always sought sustenance from the forest. Handfuls of hunter-gatherers still exist in some areas, their harmonious relationship with the environment resulting in little damage. Similarly, native farmers—often ethnic minorities—have long practiced shifting cultivation in forests. They clear a small section of land through slash and burn techniques and practice agriculture for a few years, until declining soil fertility encourages them to move on. A fallow period restores fertility and after a time the land will again support crops. Under normal conditions the damage is minimal. But these shifting cultivators are being squeezed. Land grabs and other pressures are forcing them to shorten their farming cycles and return to land they have left fallow before it is ready to support another crop. Where these pressures exist, the responsible practices of shifting cultivators are modified, and damage results.

Increasingly, the bulk of deforestation is being blamed on shifting cultivators. However, it is a new breed—whom Jack Westoby appropriately terms "shift*ed* cultivators"—that is causing this deforestation.[36] Westoby, retired director of the Programme and Operations Division of the Forestry Department at the UN Food and Agriculture Organization, states bluntly that dispossessed landless peasants, people with no hope of employment in countries with official forty percent unemployment rates, go to the forests as their last hope, and practice shifting cultivation as an alternative to starvation. But these shift*ed* cultivators rarely have the intimate knowledge of this type of farming displayed by traditional shifting cultivators. They clear lands that should never be cleared. They mine the soil and often leave it permanently unproductive. And their numbers are enormous and growing. Clearly, shifted cultivators are victims, hardly deserving the blame many place on them for deforestation. Moreover, these people are not merely the legacy of colonialism—they are often instruments of modern ranchers, clearing land for future cattle ranching operations.

Since the 1960s, the Amazon's fate has been closely tied to government policy of exploiting the hinterland. The Brazilian capital was moved to the new, more centrally located city of Brasilia (built in the shape of an airplane). And the forest was opened up with construction of the 8,680-mile trans-Amazon highway system.[37]

Inflation spurred investment in land, and the increasing demand for low quality cuts of meat for the fast food industry in the United States encouraged land clearing for ranching. To accommodate these needs, the World Bank dramatically increased its loans in the livestock sector from just four percent of its total loans between 1948 and 1960 to twenty-one percent between 1966 and 1970.[38]

The huge pool of unemployed people was putting the Brazilian military government under pressure. Since new roads gave landless peasants access to forest areas, the government encouraged people to re-settle with land grants and with payments for "improving"—clearing—the land. In the eyes of most governments, such programs are preferable to real land reform.

In Brazil, this policy has led to a wave of peasants who settle on land purely to be paid for clearing it. After a year they move onto a new plot, leaving the cleared land to a ranch or agricultural concern.

Obstreperous peasants are commonly thrown off their land, sometimes by violence, sometimes through legal means whereby corporations with political and economical clout assert their rights over land occupied by peasants.[39] According to at least one report, the Swift-Armour-King Ranch is at least partially located on Indian "reservation" lands.[40] (Under the Brazilian Constitution, Indians have not been granted the right to sue for enforcement of Brazilian laws.)

States are allowed to sell their land at a price calculated to omit the value of the timber on it. And the government gives big tax breaks to landowners clearing the legal maximum of fifty percent of their land.[41] Under a law passed in 1966, half of a corporation's tax liability "could be invested in Amazonian development projects, essentially permitting taxes to become venture capital," according to Susanna Hecht of the University of California, Los Angeles.[42]

The effect of these government policies has been to promote large-scale cattle ranching in the Amazon. Despite the "supportive" role peasants have played in this process by clearing the land—for which they are routinely blamed for deforestation—and providing labor for the ranches, their continued presence as "colonists" has drawn fire from the corporations. Ranchers claim that the peasants are unproductive and have caused damage to the environment—a problem the ranchers say they could alleviate with good corporate planning, if only they owned all the land. In effect, however, it is estimated that no more than 600,000 hectares (1.5 million acres) have been cleared by the colonists, which is "less than half the clearing . . . authorized for pasture formation [by the ranchers] in just three years, from 1974 to 1976, in the southern part of the State of Para alone."[43]

Unfortunately, the ranchers can hardly claim a clean environmental record. In 1975 the orbiting satellite Skylab, making a routine pass over the tropics, detected extreme heat originating from an area in Brazil. The information was analyzed and interpreted. Skylab had discovered a huge erupting volcano in the Amazon jungle. Urgent warnings were dispatched to the Brazilians. But when they arrived on the scene they found no volcano; they found Volkswagen burning off 25,000 hectares (62,000 acres) in preparation for setting up a huge cattle ranch.[44,45] Today on an average afternoon, literally thousands of fires—each at least a square kilometer in size—can be spotted from space. Astronauts see what we cannot: smoke from these fires covers the continent.

Volkswagen is not alone in the cattle ranching business. Standard Brands, Goodyear Tire, Armour-Dial, William Underwood Co., and Swift, among others are also in business in South and Central America.[46] In Brazil alone, three hundred ranches are raising six million head of cattle on sixty-six thousand square kilometers of forest.[47] Norman Myers estimates that thirty-eight percent of the deforestation occurring in the Brazilian Amazon during a recent period was due to cattle ranching.[48] Despite all this activity, ranching in Amazonia has been singularly unsuccessful at raising beef. In recent years only forty-four thousand dollars in beef has been exported annually from the state of Para, for example, compared to thirty-three million dollars in Brazil nuts. According to Hecht, "Amazonia itself remains a net meat importer."[49]

Converting tropical forests to pasture for the ostensible purpose of cattle ranching became a method for capturing enormous government subsidies while cashing in on rising land values. The productivity of the land and the success of the ranching operation itself were not as important as the total rate of return on investment. The commodity was not beef, but land. Enormous profits could be made in Amazonia even in the face of deforestation and failed cattle ranching.

Central American cattle raising got its impetus in the 1960s with the explosion of fast-food hamburger restaurants in the United States. Burger King (the third largest fast-food chain in the U.S.), Roy Rogers, and Hot Shoppes acknowledge that they use beef imported from Central America.[50] In truth, most chains do, though some, including McDonald's, deny it. It has been reported that one McDonald's supplier, Industría de Ganaderos Guatemalecos SA, concedes that they obtain their beef from Guatemala.[51] After imported beef is inspected at a U.S. port of entry, it no longer bears the label "imported." This bureaucratic maneuver changes Central American beef into domestic meat with one stroke of a purple U.S. Department of Agriculture stamp.[52] Meat can sometimes change ownership several times while lying in a warehouse, before it becomes a "Big Mac" or a "Whopper."[53] But the connection between the North American hamburger industry and tropical deforestation is unquestionable.

Cattle ranching on tropical forest land is no simple matter. By the time cattle are munching away on the grass, the land may have been logged by a timber company, supported a few years of agricultural crops grown by a peasant-settler, or both. Stocking rates begin low—one animal per hectare—and get worse, dropping to one or two cattle per five to ten hectares.[54] But as long as there is plenty of forest to be cut, and government incentives to do so persist, there is no need to look for greener pastures.

As U.S. beef imports have increased, per person consumption of meat in Central America has declined to a point below that of a domestic cat in the U.S.[55] Using land to raise beef and grow export crops has deprived thousands of people of their means of making a living and feeding themselves. The poor would lack the money to buy the meat even were it not exported. The meat follows the money.

Beef, big business, and politics are the ingredients of the Central American stew. Nicaragua's former dictator, the late Anastasio Somoza, for example, held interests in six beef importing companies in Florida and raised cattle on fifty-one haciendas in Nicaragua (to say nothing of his forty-six coffee plantations). Until the revolution in 1979, the U.S. imported more beef from that country than from any other in Central America.[56] Costa Rica now enjoys that honor. But Costa Rica's pastures have been suffering from drought, due to decreased rainfall and increases in the amount of runoff caused by deforestation. The deforestation is also causing severe soil erosion, which according to Norman Myers "has caused dams to silt up, in turn bringing shortages of drinking water and electricity."[57]

Paul and Anne Ehrlich of Stanford University contend that "more than a quarter of all Central American forests have been destroyed in the past twenty years to produce beef for the United States"[58]—an activity supported and financed by billions of dollars of World Bank and U.S. "aid" money, and millions more in technological assistance from the U.S. Department of Agriculture, Organization of American States, and the Pan American World Health Organization.[59] Today some two-thirds of the region's arable land is devoted to cattle production.[60]

Meanwhile, Washington ponders the Soviet and Cuban roles in revolution. We would do better to look closer to home. As James Nations and Daniel Komer observe in an incisive report on forests and cattle ranching: "Americans must be made aware that when they bite into a fast-food hamburger or feed their dogs, they may also be consuming toucans, tapirs, and tropical rainforests."[61] Nations and Komer give a whole new meaning to the McDonald's slogan, "We do it all for you!"

Timber companies are not the villains they are sometimes made out to be in tropical deforestation, but neither do they win any prizes as model citizens. Third World governments rarely receive adequate compensation for the timber that is taken. Bribery of local officials is standard operating procedure. Aside from a few showcase examples, reforestation and conservation efforts are totally inadequate. The big timber companies find tropical forests inviting because the wood they cut there is cheaper than anything else they can get. And it is cheaper precisely because of the practices mentioned above.

WHO'S TAKING THE FORESTS?

In some areas these practices have drawn opposition. Between the Eia and Mambara rivers in Papua New Guinea lies the land of the Binandere tribe. Five thousand people in this tribe live in what we might call "Stone Age" fashion. But they have done something remarkable. They have learned to live in balance and harmony with the rain forest. During the last ten years, preserving that life has meant preventing an American-owned timber company, Parsons & Whitmore, from cutting down the forest.

Addressing a Swedish documentary television crew through an interpreter, the Binandere chief, Kipling Jiregari, spoke with great emotion:

> My name is Kipling Jiregari and I shall speak of the land and of the forest. When God created the world and Papua New Guinea, he created our country . . . Here we all in the Binandere tribe have tended our gardens. This land gives us our food and everything we need.
>
> Money has no future. Money disappears. Only Man and the land remain, when all else has disappeared. Therefore I have stopped the Company. No company shall destroy our land.
>
> If the Company's men come here again, we will kill them and eat them up. We received our land from God and one day He will ask to get it back . . . If one of the Company's men comes here, I'll run him through with my spear.
>
> They shall never touch our wood, where our food and our medicine is. The wood is our skin, and without his skin, Man dies . . . The woods, the land, the fish, the swamp, the birds—everything we will

> protect—not only for our own sake, but also for coming generations . . .
>
> We don't want their money—tell them that. Our ancestors did not live on money; we are not descended from money. We descend from the taro plant which our ancestors lived on—and which we live on. We are made of taro, not of money.
>
> We did not ask the Company to come here. We have all we need. The woods and the land are ours. No one shall come and try to bargain away our woods—let me hear no more about wood-cutting rights.
>
> The Government can invite the foreign companies to its own areas. But they shall not come to the land of Binanderes. The forest is our skin. Take away a person's skin and he dies.[62]

And so it is. The Binanderes are lucky. For the time being they have saved their skins.

Four hundred and fifty kilometers away the native people living in a forest granted to JANT, a subsidiary of the huge Japanese multinational corporation Honshu Paper, are not so lucky. JANT seems to subscribe to the Montana loggers' slogan that "the only good tree is a stump."[63] Formerly, timber companies selected for cutting only those trees with valuable commercial uses—the mahogany and teak. Honshu, however, became the first company to clear-cut the tropical forest and turn it into wood chips.

It takes a worker just seventy seconds to fell one of the giant trees. As it crashes to the ground, smaller surrounding trees and those connected to the giant by vines in the canopy come down too. The trees are hauled to JANT's plant in Medang, where fifteen tons of steel pressed into eight huge blades revolve hundreds of times a minute, and spit out the forest in the form of billions of tiny wood chips. Later at factories in Japan the chips will be pressed into cardboard boxes or made into toilet paper.

The machines have destroyed this forest. Springs and water holes have dried up. Birds and wild game have left. What remain are the people, whose sustenance for as long as they can remember has come from the forest. A man from the village speaks:

> The Company came and took our forest. We said okay, because we thought it was good. But now they have taken our forest, and things have not gotten better for it . . . Now when the forest is gone both the Government and JANT are gone too.

> What shall we do now? We cannot live without food from the forest and from our gardens . . . How shall we make them understand—that now, when we do not have the forest, we can no longer look after ourselves?
>
> If we are to eat, we must now have shops where we can buy food—and money to pay with. But we have neither shops nor money.[64]

In recent years, demand for tropical hardwoods in world trade has grown at a dizzying pace. In 1950, the producing countries in the tropics consumed five times more tropical industrial timber than Japan, the U.S., and Europe. But today, consumption levels are about even.[65] Between 1950 and 1973, Japan increased its imports of tropical hardwoods by nearly two thousand percent,[66] the U.S. by almost a thousand percent.[67]

The loggers vary. Many are one-man, fly-by-night operations, timbering anywhere they can get away with it, including national parks. Some are familiar multinational timber corporations. Of these, the U.S.–based are most active in South and Central America. Four (Georgia-Pacific, Olin-Kraft, Scott, and Westvaco) presently control over 1.5 million acres in Brazil alone. Japanese corporations dominate the Pacific area, though Weyerhauser has recently held rights to a 1.75-million-acre concession in Indonesia.[68] And European companies are active in Africa and South America.

In South and Central America it is often difficult to separate timbering from other causes of deforestation. Centuries of colonialism and one-crop economies left many of the nations destitute, dependent, and hungry. By the sixteenth century, the Dutch, English, French, Spanish, and Portuguese were vying to control the Brazilian coast. Already the area was valued for its sugarcane,[69] a crop from Southeast Asia. Indians were driven from their homes in wars with the Europeans. As deadly as the wars and enslavement were diseases, like measles and smallpox, brought by the Europeans. Lacking immunity, whole Indian societies were wiped out.

Finally, we would like to mention one other cause of deforestation, rarely noted in academic works on the subject: war. The term "ecocide" entered the English language during the Vietnam War, perhaps because destruction of the environment played a crucial role in the strategy and tactics of the United States.[70]

Between 1965 and 1971, Indochina—an area only slightly larger than Texas—absorbed twice the tonnage of munitions that the U.S. used in World War II.[71] Twenty-six billion pounds fell on Indochina—the equivalent of 450 Hiroshima bombs, 142 pounds for every acre of land, 584 pounds for every person.[72]

The typical five-hundred-pound bomb (108 of which could be carried by a B-52) left a crater fifteen feet deep and thirty feet across. There are twenty-six million of these craters in Indochina, covering 423,000 acres.[73] Aside from simply "wasting" large areas of Indochina, shrapnel from the bombs left cuts and gashes in millions of acres of forest vegetation. In the tropical climate, diseases attack and spread quickly when given such an opportunity. During the final years of the war the U.S. employed a particularly destructive bomb capable of completely clearing out a 1.3-hectare area for use as a helicopter landing pad. The "casualty" zone of the bomb was a staggering forty-nine hectares.[74] As a result of these and other bombs, one mill owner in Vietnam claimed that four out of five logs used by his mill came with metal in them.[75]

Some two percent of the land area of South Vietnam was cleared by "Rome ploughs," [a huge chain pulled by two tractors or bulldozers][76] including a fifteen hundred-acre area which was carved out in the shape of the emblem of the U.S. 1st Infantry Division, twenty-five miles from Saigon.[77]

And then there were the chemicals. Thirteen to fourteen million pounds of 2,4,5–T—the entire U.S. production in 1967 and 1978—were used by the U.S. Department of Defense. This amount is sufficient to obliterate the vegetation on ten million acres if used properly.[78] All told, fifty-five million kilograms of herbicides were used.[79] Herbicides were directed not only at tropical forests but also at rubber plantations and rice paddies. Many of these areas were sprayed repeatedly.[80] No estimate is available of the number of traditional and locally adapted rice varieties lost as a result of the war, but the number is surely high. At the end of the war, Dr. David Ehrenfeld of the Department of Horticulture and Forestry at Rutgers University said: "To say that every ecosystem in Vietnam, major and minor, has been seriously altered or wrecked beyond hope of repair, is to make a safe and conservative statement."[81]

VALUE OF THE FORESTS

The tropics cover twelve percent of the land area of the world. Their forests cover four percent,[82] but they probably contain the majority of the world's species. Is it likely that these forests can be destroyed without ill effects for the human race? Hardly.

Many appeals to save the tropical forests come from well-meaning people concerned about the fate of butterflies and ferns. This is well and good, and in the long run it even makes sense economically. We would also argue that it makes sense environmentally, and that indeed, these forms of life have a basic right to exist. But, as Richard Nixon was fond of asking before making a political decision, "Will it play in Peoria?" The answer, unfortunately, is no. So let us look instead at something with more immediate implications for most of us: the morning cup of coffee.

Among the millions of species native to tropical forests are rubber, cacao, cassava, cashew, and vanilla; pineapple, pomegranate, and numerous other fruits; trees like ebony and teak; numerous drugs used to fight diseases including cancer and malaria; and, of course, coffee.

Evidence from a UN Food and Agriculture Organization (FAO) mission to Ethiopia in the mid-1960s has established the rain forests of southwest Ethiopia as the place of origin of *Coffea arabica*.[83,84] Outside Ethiopia, there is little genetic diversity in this crop. Virtually every coffee tree in South America traces its heritage to a few seedlings from a single tree found in the Amsterdam botanical gardens more than two hundred years ago, making the genetic base of the coffee industry about as narrow as it could be.[85] More frightening yet, a virulent new disease is striking coffee. Little resistance to it has been found, due in part perhaps to the fact that seven-eighths of the forest in Ethiopia had vanished by 1965,[86] creating the conditions for what the International Board for Plant Genetic Resources (IBPGR) terms "disastrous genetical erosion."[87]

Added to this problem is a political complication. Convinced that their genetic resources have benefited just about everyone but themselves in the past, and lacking guarantees that this will not happen in the future, the Ethiopians have closed their doors to future collection of coffee resources in their forests. Our discussions with the minister of agriculture have left us with the opinion that the Ethiopian position is firm. Unless Ethiopia relents, we will have an opportunity to see what happens to a narrowly-based crop like coffee without recourse to badly needed genetic resources. Our advice to coffee addicts? Have you really given tea a chance?

In recent years, bananas have been faced with serious disease problems.[88] They must be sprayed ten to thirteen times in the Americas and even then, disease control is not assured.[89] There is some question as to the safety of banana collections in Lae, New Guinea, in the center of greatest diversity for bananas.[90] And the IBPGR cites "evidence of widespread loss of cultivars in the centre of diversity . . ."[91]

When last found in 1967, *Punica protopunica*, the pomegranate's only wild relative, had been "reduced to four aging trees on the island of Socotra" in the Indian Ocean. Only twelve *Persea theobromifolia* trees are left in a reserve in Ecuador. Not only was this tree once an important timber tree, but it is also a near relative of the avocado that "could be of great breeding value in the future" according to the FAO.[92]

Lovers of cakes and vanilla ice cream may live to mourn what is happening to the important orchid which the Spanish conqueror Cortez found the Aztecs using to make a drink. Carl Withner of Brooklyn College tells us that in Mexico, Papantla was the main vanilla-growing center. "But, sad to say, when I visited Papantla about 15 years ago, the only vanilla I could find was in a tile painting on the back street . . . in the old section of town." Oil had been found nearby, vanilla substitutes had been developed, and the vanilla growing habitats had been sacrificed.[93]

A recent study by our colleague Hope Shand of the *Rural Advancement Fund International* has uncovered a new threat to vanilla and those who make their living from raising it: biotechnology. Two companies in the U.S. are predicting that they will soon begin commercial production of vanilla in the laboratory. This will not be a substitute, but the real thing produced without the need for the plant or the farmer. Over seventy-five thousand farmers, mainly in Madagascar, may soon find the market for their crop disappearing.[94]

Cassava is a crop rarely thought of by people outside the tropics, but in the tropics it is the staple food of more than two hundred million people. It is beset by diseases which have provoked regulations prohibiting the exchange of cassava between continents for fear of introducing the diseases into new areas.[95] While not yet urgent, collection and preservation of cassava genetic resources is surely a priority.

Various tree species are threatened both in and out of the tropics. A once common relative of the English elm that shows resistance to Dutch elm disease is now considered an endangered species in the Himalayas. The tree might be of use in trying to overcome this destructive disease. But on a recent trip, a Dutch botanist and elm specialist could find just three flowering specimens.[96]

In Honduras and much of Panama, mahogany has become extinct. Teak is disappearing in Southeast Asia.[97] *Cinchona officinalis*, the tree whose bark is the source of the anti-malaria drug, quinine, is endangered in its home in the Ecuadorean cloud forests.[98] The number of timber trees endangered is too great to enumerate. One tree which accounted for sixty percent of Nigeria's timber exports in 1970 is no longer exported at all. "This reflects not the disappearance of the species," according to a UN report, "but the loss of its economic potential through loss of genetic diversity."[99]

Scientists concede that the genetic resources of trees—once thought to be relatively secure—are increasingly endangered. Christel Palmberg, forestry genetic resources expert for the FAO, is not given to overstatement and is critical of those who are. But the case she makes—verbally and in publications—is ominous: "Every one of the tropical/sub-tropical forest tree species explored and collected during the present decade in a programme coordinated by the FAO has, as a result of the exploration, been found to be in danger of depletion, extinction, or contamination of its genetic resources in at least part of its natural range . . ."[100]

FAO officials are normally loath to admit that new varieties of agricultural crops are replacing the old and driving traditional varieties into extinction. But an FAO/UN Environment Program panel has cited yet one more, regrettably familiar cause of genetic erosion of forest species. "It is only realistic," they write, "to assume that an expansion in breeding (of modern tree varieties) coupled with the depletion of natural forest ecosystems will create a situation of genetic erosion similar to that experienced in crop plants . . ."[101]

Although the tropical forest contains species of obvious, direct economic benefit to humanity, it is the forests' effect on the environment that may soon provide the most compelling argument for their preservation. Tropical rain forest provides an efficient way of utilizing vast quantities of water. More than three times as much water falls per acre in the equatorial rain belt as the global average.[102] Much of this water comes from transpiration. Water discharged from the forest creates massive river systems. The Amazon River has a thousand major tributaries and twenty thousand kilometers of navigable waterways, "two-thirds of all river water of the world . . ."[103]

With the forest intact, water is processed efficiently. Remove the forest, and floods and soil erosion result. Deforestation causes a hotter, drier climate, more prone to drought; at the same time the soil is baked in the sun and loses its water retention capabilities. Already, rising flood crests are evident in the Amazon.[104] And an increasing number of deforestation-caused floods are being reported from tropical regions around the world. In parts of Asia, rising flood waters are literally drowning new short-stemmed rice varieties.[105]

As the water runs off the hardened soil to swell the rivers, it takes with it such topsoil as the forest has. Soil losses in a virgin rain forest are negligible, but left bare or converted to field crops, land in the Ivory Coast annually lost 90–138 tons of soil per hectare. In extreme cases, losses of up to twelve hundred tons a year have been reported.[106] Some of the soil is blown away—dust levels in the lower atmosphere have tripled in the last sixty years.[107] But most ends up in streams and rivers. In the Philippines, one major river carries away an average of 44–46 tons of soil yearly for each hectare in its watershed. The "muddy" Mississippi River, by contrast, carries only half a ton per watershed hectare.[108]

River-borne soil silts up reservoirs, decreases hydroelectric output, and threatens supplies of drinking water. Left unattended it will even prevent large ships from passing through the Panama Canal someday.

As regards environmental integrity, the good news about tropical forests is that all the positive functions they perform—from conservation of plant genetic resources to conservation of soil—are free. The bad news is that destroying tropical forests carries enormous costs, chiefly in agricultural productivity. In 1981, Robert and Christine Prescott-Allen surveyed agencies charged with administering protected areas like parks and reserves in fifty countries. Fewer than half the respondents had catalogued the species to be found in any of their protected areas. The Prescott-Allens concluded that "it is impossible to tell how many wild relatives of crops occur in protected areas; hence to decide whether new areas are needed and, if so, where." Not knowing what is in the reserves would make it hard to use material there in plant breeding programs. But only fifteen percent of the agencies would allow collection of reproductive material anyway.[109] Both destruction of tropical forests and certain conservation practices have their costs. They may even lead someday to the end of commercial production of some crops, thus increasing hunger.

Were these costs not enough, there is one more, often overlooked. People.

The world's tropical forests are populated, if sparingly, by quite a few people who could not tell you the difference between Paris and a bottle of Coca-Cola. But many of them could tell us a good deal about plants.

The hunting and gathering Hanunoo of the Philippines can distinguish sixteen hundred categories of plants.[110] In Thailand residents of one forest village eat 295 different plants and use a further 119 for medicine. Worldwide, the World Health Organization estimates that three thousand plant species have been employed for birth control by tribal peoples.[111]

Such matters are not lost on the drug industry. A recent study warned that some two hundred drug-yielding species were in danger of being lost. Referring to an article in *The Guardian,* the study suggested that the industry could lose well over $100 billion in prescription medicine value. The price tag on the extinction of each medicinal species was estimated at $203 million.[112]

At a time when we in the North yearn for drugs to cure or obscure our many ills, the world's only real herbal medicine experts—the native peoples of the South—are dying out. And even when their knowledge of the curative powers of plants is saved, it rarely if ever yields profits for them. In 1500 an estimated six to nine million Indians lived in the Amazon region. Wars and diseases took their toll. By 1900 there were only one million[113] in 260 tribes.[114] Today there are perhaps two hundred thousand. During this century eighty-seven tribes have disappeared;[115] twenty-six have become extinct during the past decade.[116]

Let there be no mistake about it, when we say they "disappeared" or became "extinct," we mean they were killed or scattered. Their rich, long cultures ended; their languages are dead, never again to be heard. With this destruction, we lose knowledge. We forfeit the specialized knowledge these people gained from hundreds or thousands of years as conservationists amongst trees. Some forest dwellers have also been both nonviolent (lacking words for violence, war, and the like) and democratic (practicing collective decision making). We have thus lost the opportunity of learning more than just plant-lore from them. Lessons about how to get along with one's neighbors have gone unlearned.

Further tribal extinctions are to be expected. Some of the richest iron ore deposits in the world are located in the Amazon Basin. Bauxite and gold are also there. Laws passed by the Brazilian Congress in 1973 institutionalized the Indians' status as tenants by declaring that all of the subsoil riches of the region belonged to the government to exploit as it pleased, without need even for compensating the Indian population.[117] Mauricio Rangel Reis, the Brazilian minister of the interior, got straight to the point: "The ideals of preserving the Indian population within its own habitat," he observed calmly, "are very beautiful ideas but unrealistic."[118] Toward that end, the Brazilian Indian Foundation, the government agency that administers Indian lands, offers an even more cynical view. The agency's policy, according to its director, is that "Indian programmes shall obstruct neither national development nor the axes of penetration for the integration of Amazonia."[119]

In a classic example of even more blatant insensitivity, C.W. Bingham of the giant U.S. timber company Weyerhauser was able to look at this sordid history and state that "changing the local society . . . involves cultural enrichment . . . I feel no apologies are necessary for offering members of these societies a choice."[120] What we have here is one of those classic "failures to communicate," a misunderstanding of huge proportions. One wonders how the Binanderes of Papua New Guinea would have gotten their point across to Mr. Bingham!

DEMOCRACY AND TROPICAL CONSERVATION

Time was when conservation regulations were pretty simple. In the Garden of Eden there was one rule. The only two people who could have violated it quickly did.

The Roman ruler Hadrian sought to remove the freedom of choice element from forest conservation in 125 A.D. by putting the authority of the Roman empire behind his decrees—chiseled in stones on the Lebanese hillsides—that the cedars of Lebanon not be cut. Today, the stones are about all that is left protruding here and there from the barren hillsides. The trees were cut by the Egyptians, Phoenecians, Chaldeans, the Romans, kings David and Solomon for their temple, and even the Allied Army during World War I. The best stand of cedars occupies less than twenty acres today.[121,122,123] There are more cedars on Lebanon's flag and postage stamps than rooted in the soil.

Rules and refuges are likely to be about as effective in the future as they have been in the past unless the real causes of deforestation are addressed. We have seen the political and economic nature of those causes: the legacy of colonialism; the tight concentration of land, wealth, and power resulting in peasants fleeing to forests to avoid unemployment and starvation; and damage caused by foreign-dominated, export-oriented timber and cattle businesses.

Given this level of exploitation, is it any wonder that movements for social justice, insurrections, coups, and revolutions are commonplace? If the real causes of deforestation are to be dealt with, some of these uprisings will have to succeed in bringing land reform and a degree of democracy to oppressive governments and economies. Yet today many popular movements incur the instant opposition of the United States government and those of other Western powers. Guns, advisors, and Central Intelligence Agency functionaries are routinely supplied to the most despotic regimes if they are threatened by calls for land reform or true democracy. A change in U.S. foreign and military policy is essential if Third World peoples are to solve the problems of hunger and jobs without resorting to cutting down tropical forests.

Swapping foreign debt for the setting aside or protecting of reserves is a new and positive initiative on the part of governments and lending agencies. It should be encouraged. But, unless the lowered debt actually translates into improvements in the lives and futures of the landless and the oppressed, it will not ultimately be able to save the forests.

It would be consoling if the problem of deforestation could be solved with parks and reserves. Presently these sanctuaries occupy little more than one percent of the world's land area, however, and much of the conserved land is in Greenland and the Canadian Arctic, where the fewest species are located.[124]

While countries in the tropics have designated a number of reserves, most exist on paper only. There being little money for maintenance, the reserves are patrolled poorly, if at all. Some parks have human populations of squatters and farmers that number in the thousands. Others are crisscrossed by logging roads. One study puts the amount of land "effectively protected" in the Amazon Basin at one-quarter of one percent.[125] Incredibly, some park sites seem to have been chosen not because of their biological significance, but "on the basis of low potential for other uses with direct economic benefits such as mining, forestry, or agriculture."[126]

Two questions arise: Where are the species in need of protection? How many individuals are needed to carry on the species?

In the Amazon, we know of the existence of at least sixteen areas with extremely high concentrations of species. Some scientists believe these are areas which maintained their diversity through the long dry period during the Pleistocene, and thus enabled many species to recolonize the Amazon afterwards. These areas constitute the heart and blood of the Amazon. Unfortunately, four lie completely within regions designated for the development of agriculture and cattle ranching.[127] At the very least, all sixteen of these presumed "Pleistocene refuges" must be protected. In order to determine the areas most suitable for conservation in other tropical forests, much more research will have to be done.

The second question—how many individuals are needed to carry on the species?—is more complicated. Extinction occurs long before the last individual of a species dies. As long as the world changes, the ability to evolve and adapt to those changes will be a precondition of survival. Ability to evolve depends on genetic diversity. Once this diversity is lost, the ability to evolve is lost—and the species is lost. One must understand that extinction is a process, not an event like the death of the last dodo. With fewer than two dozen captive individuals left, the huge California condor, for example, is both very much alive and very much extinct.

What is the minimum number needed? Scientists debate that question. Their answers range from two hundred to ten thousand.[128] Sir Otto Frankel and Michael Soule, in a book not quite old enough to be a classic, argue that the minimum number of breeding individuals for each species, equally divided between male and female, is five hundred. In order to obtain that many fit, breeding-age animals, the survival of perhaps a thousand or more individuals would be necessary.[129] Frankel and Soule's estimate falls near the low end of the range and thus for the purpose of judging the adequacy of present-day reserves, might be considered conservative. Moreover, recent studies have cast doubt on the methods used in arriving at some of these estimates,[130] indicating that we still are not sure how many is enough. Be that as it may, the two scientists note that only 3.5 percent of the world's national parks exceed ten thousand square kilometers, and argue that "even this size is much too small to maintain viable populations of the largest carnivores . . ." They conclude that "without intensive management, it is likely that the majority of birds and large (greater than one kg) mammals will be extinct in a few thousand years and that these extinctions will precipitate complex chain reactions leading to many other extinctions in all taxa."[131]

When will this process begin? Why, of course, it already has! Soule estimates that the largest reserves (themselves too small to protect diversity and ensure evolution[132]) may see a loss of thirty percent of their large mammals in the next five hundred years.[133]

Our knowledge of tropical forest ecosystems is still rudimentary. It has been said, not entirely in jest, that we know more about the far side of the moon than we do about tropical moist forests. In the years to come, scientists will learn more, and some of that may come from the current experts—the indigenous hunter-gatherers and the shifting cultivators.

But more research and knowledge and the most elaborate system of parks and reserves will not achieve conservation ends if fundamental changes are not made in society itself. "Fences and guns will not protect forests against hunger," Jack Westoby observes. Only political changes and the active participation and support of local people can do that. Without those changes, no conservation program will succeed, however well planned.

PART TWO

Genetic Technology and Politics

CHAPTER SIX

Rise of the Genetics Supply Industry

The American Beauty Rose can be produced in its splendor and fragrance . . . only by sacrificing the early buds which grow up around it.

John D. Rockefeller, Sr.

When American scientists with the U.S. National Academy of Sciences studied the genetic vulnerability of major crops, they concluded that "powerful economic and legislative forces" contributed to the uniformity of the food system.[1] Over the past three decades, the private seed industry has taken charge of the distribution of "high response" seeds in the North and is now turning its attention to the South and the centers of genetic diversity.

Even as the seed trade expands, it is being transformed into a "genetics supply" industry dominated by the transnational enterprises that manufacture farm chemicals. An industry that traces its origins back to the travelling tinkers and hucksters of two centuries ago is now tinkering with the first link in the food chain. The survival of genetic diversity may now be in the hands of that industry.

Wall Street analysts study the entrails of the chemicals industry, dig deep into the bowels of mining and manufacturing houses and follow the tangled trail of the microprocessors from Silicon Valley to Singapore. Chicago's commodity brokers know more than anybody ever wanted to know about pork bellies. In London, Mincing Lane's auction traders study tea bags—if not tea leaves—while their confreres chart the course of September wheat by the vessel hirings at the Baltic Exchange. But where are the analysts for genes—the stuff that gives those Baltic ships their ballast, fills pork bellies, and puts the orange in Orange Pekoe?

The advent of biotechnology companies and the public offering of shares in the next cure for cancer—or, at least, hoof and mouth disease—has drawn the enthusiastic interest of investment houses. For the most part analysts monitor boardrooms and technology. But what about the raw materials supply—the seeds and genes? For thirty thousand dollars a small Milwaukee company (the refuge of a former seedsman) will hand-deliver a copy of their analysis of the impact of genetic engineering on the seed industry.[2] Even in hardcover, the edition is overpriced. Other studies, such as those sponsored by the French, Australian, Canadian, and American governments, tend to be apologias for past policies or proposed legislation rather than actual evaluations of the status of genetic raw materials. The food system's most precious and precarious raw material is stored, stolen, swapped and sold almost entirely out of sight of the farmers and consumers who depend upon it.

Although statisticians can tell us not only what we eat but where, how much, and at what cost, they have little to say about seeds. Information on what seed varieties are grown in what quantity, and who the breeders, retailers and growers are, is only vaguely known. As the first link in the food chain, seed is a central factor in any nation's strategy for food security and self-sufficiency.

TRAVELS OF A SEED

The metamorphosis of the seed industry into the more exotic genetics supply industry is almost painfully complete. The bright new butterfly of the venture capitalists still includes the enterprises that breed new varieties; companies that grow or contract for the growing of new seed; international seed brokers and discard hucksters (for "old" seed); seed cleaners and conditioners; and the wide assortment of wholesalers and retailers from big mail-order houses to corner grocers. But today the industry also includes seed pathologists, plant physiologists, soil scientists, plant breeders, geneticists, microbiologists, toxicologists and, now and again, a lonely nutritionist. Some seed houses work exclusively with vegetables and herbs. Others deal in forage and cereal crops. Their final product may eventually reach the public through farmer/dealers, catalogues, flower shops, or farmer-oriented merchant supply stores.

En route to the cabbage patch, a new variety may begin with a collecting expedition to the Near East funded by a quasi-United Nations agency in Rome. Germplasm from the expedition may be evaluated at a government facility in Warwick, England. Combined with traditional local cultivars, the improved material will be made available to a private breeding concern at King's Lynn in the United Kingdom, which will pass it on to a vegetable-breeding sister company in Enkhuizen, Holland.

Taking advantage of differing growing seasons, the Dutch concern may contract some breeding work to a partner in Christchurch, New Zealand, or Santiago, Chile. Ultimately, the new cabbage may be multiplied (this involves growing and regrowing seed until sufficient quantities are available for commercialization) by another sister firm in Arusha, Tanzania. Cheap land and labor, a favorable climate, and an absence of diseases conspire to make Arusha one of the world's largest growers of vegetable seed, and a home away from home to a dozen international seed companies.[3]

Ready for market, the new variety will be licensed by its owners to wholesalers and retailers. Through the services of seed brokers, this "Dutch" cabbage may appear in the catalogue of a Brandon, Canada, mail-order seed house or be sold in grocery stores in Seoul, Korea. Some of it may even be marketed back to gardeners in the Near East.

Agricultural seed (such as cereals and forages) may develop along a roughly parallel route with a few additional kinks and twists. In most industrialized countries (though not in the U.S.), the new variety will have to undergo a battery of field tests intended to see whether it is an actual improvement over other varieties. If it survives, the approved seed will be entered on the "national list" or, in the European Economic Community (EEC), in the "common catalogue." Those that don't make the grade are doomed to the dustbin or destined to become a new "wonder" seed dumped on a Third World country.

These days, the well-groomed farm seed comes wrapped in a clay mudpack and gussied up in chemicals. After the seed has been sold and planted, its offspring may be hauled back into town the following year to be "cleaned" and even "fumigated" before being dropped back into the seed drill for another bout with nature.

For all the oft-told joys of genetic engineering, it may be the seed clothiers—the plant growth regulators, coaters, and inoculators—that will make up the boom side of the genetics supply industry. Passing are the days when seed could be saved and, come spring, be allowed to venture into the world naked and alone.

The annual retail value of the world's seeds in the mid-1980s may border on $51 billion.[4] No one knows for sure. Since this figure includes the nursery trade, farmer-to-farmer exchange, saved seed and heavily subsidized government seed distribution, the actual commercial value of sales is closer to $32 billion.[5] But the gene business is pivotal for the $17.4-billion pesticides industry.[6] Beyond this, seeds are obviously the central factor in the multi-trillion dollar food industry—the largest and most important enterprise in the world.

Profit may prove a clearer indicator of seed value. One informal survey taken by the antitrust officials at the EEC in Brussels suggests that a full disclosure of the industry's return on all aspects of the seed trade—including seed multiplication and licensing—would reveal a forty to forty-five percent return on sales, an awesome profit by any standards.[7]

In a 1986 survey listing seed sources for 150 countries, the UN Food and Agriculture Organization (FAO) identified nearly seven thousand public and private sources offering planting material for about 250 crops.[8] The OECD (Organization of Economic Cooperation and Development) lists 743 public and private seed houses from the thirty-six countries participating in its international varietal certification schemes for cereals and forages.[9] Because many seed companies are small, family-run operations, we may never have a true figure of the scope of the industry. A wise guess would place the total number of active breeding and/or marketing enterprises (thus excluding general retail supply stores, flower shops, etc.) at substantially over two thousand, of which more than three-quarters are in the western industrialized countries.[10]

ALCHEMY

The roots of the genetics supply industry are threefold. One set—if not the earliest, certainly the strongest and most productive—is found in the public domain: the network of botanical gardens, colleges, government research stations, and international research centers that still gives the industry many of its true innovations. A second set—arguably older—makes up much of the private sector today. The third source of genetic supply for agriculture sprang from the farmers themselves, obviously. The cooperative movement has been a breeder of new varieties and continues to be a major conduit of seed to farmers around the world.

In the modern world, breeding and distributing seeds has been the task of national governments, government-funded universities and institutes, and a network of nonprofit, semi-governmental international research stations. The extensive role of public agencies has known no ideological boundaries and is as much a part of the farm history of the United States as it has been in Sweden or the Soviet Union. Governments have long recognized that farming is both a business and a way of life, and that the family farm cannot be expected to bear the full burden of agricultural research when the benefits of that research accrue to the whole of society. Of all industries, farming is the least able to pass along research and development costs to the rest of the community.

With an annual budget of around $230 million[11]—most but not all of which is devoted to breeding or related agronomic research—the International Agricultural Research Centers ("IARCs") play a central role in developing germplasm and distributing samples to other institutions in almost every country. The IARCs are leaders in the development of corn, wheat, barley, sorghum, millets, pulses of all kinds (leguminous plants such as peas, beans, and lentils), cassava and rice. In dollar terms, this is surprising since their total budget approximates funding for only four leading U.S. university research programs.[12] In the context of Third World agricultural research, however, the amounts are large and, combined with the flexibility and authority provided by their quasi-UN mandate, the IARCs are pivotal in setting the direction of agriculture in the South.

The most famous of the IARCs is Mexico's CIMMYT, the International Maize and Wheat Improvement Center that gave birth to the green revolution. Newer, almost as famous, and arguably more prestigious today, is the International Rice Research Institute in the Philippines—IRRI. Collectively, the IARC network represents the elite frontline for the spread of western agricultural technology to the Third World. The IARCs also serve a major role in the collection of Third World germplasm and its diffusion to industrialized countries.

But the IARCs are a phenomenon of the 1970s, whereas the world's leading national breeding institutions are a phenomenon of the 1870s.

Some claim that the agricultural university in the suburbs of Vienna is the oldest plant breeding institute in the world. If they are right, it is not by many years. The first state experiment station in the U.S. was established in Middletown, Connecticut, in 1875,[13] long before the Plant Breeding Institute at Cambridge, England, officially opened its doors in 1912.

If we look beyond agronomic seed (field seed for farmers), then the history of public research into plant improvement goes back much further. The old Amsterdam Botanical Garden recently marked its three hundredth anniversary and Kew Gardens are older than the Vienna university by more than a century. Not long after the Austrian monk Gregor Mendel had completed his work on peas and written up his findings on heredity in the 1860s, an international network of publicly funded research institutes was in place. Mendel's work, the first to lay out our own modern theories of heredity, went unnoticed even as Darwin's created a stir. But in 1900, three scientists found Mendel's studies.

In the days before the rediscovery of Mendel's Laws and the beginnings of genetic research, the furnishing of new seed was half farmer skill and half trader alchemy. Much of this early history is notorious. Carl Buchting of the West German seed company KWS speaks of those early days as a time of thieves and shysters, peddlers who crossed Europe hawking bags of "miracle" seed.[14] Far from miracles, the bags were often filled with weed seed and pebbles.

There is another view. Cliff Swartz, president of a Canadian seed company and grandson of a Russian seedsman, describes his business as one of the most honorable in the world. It is a view—not surprisingly—that many in the seed business share. The seed trade, according to this view, is rooted in the landed gentry and aristocracy of nineteenth century Europe. Gentlemen-farmers undertook to improve seed varieties and to clean their seed before replanting. Either from hobby or husbandry, the squires began to sell their seeds and services to neighbors.

The seed trade has yet other roots. On a summer's day Paris tourists flood along the banks of the Seine making their way from Notre Dame to the Louvre. On their way to marvel at the treasure trove captured by Napoleon during his Egyptian and Italian campaigns, and the vast wonders surrendered by former colonies from India to the Ivory Coast, most tourists grant little notice to another artifact from the same period—an ornate, rambling, three-storey edifice bearing the sign, "Vilmorin-Andrieux." The Continent's oldest seed house is hidden by its own sidewalk stalls, including a great arcade of potted plants, seed packets, and cages of exotic fowl.

The traditions of the Vilmorin-Andrieux seed firm date back to 1727 when the old house followed French troops about the globe collecting exotic new plants for sale to the aristocracy of pre-Revolution France. Cuthberts of England can claim a yet longer history, dating back at least to the last quarter of the previous century, when the firm imported exotic camellias and azaleas for the stately homes and gardens of the British gentry. In the halcyon days of Queen Victoria, seed (or plant) catalogues had hundreds of pages and were crammed with varieties and whole species rarely seen in today's more pedestrian catalogues. In fact, up until the end of World War I, these merchants of exotica *were* the seed trade.

The depression after the war changed all this. The "stately homes" ceased being so stately. Many of the great merchant houses followed their plants into seed and the rest drastically reduced their offerings. Garden vegetables replaced the more delicate ornamentals.

Companies that survived did so only to find themselves attractive targets for takeover by larger enterprises. Cuthberts was more than three hundred years old and had its own listing on the London Stock Exchange when it was gathered into the corporate fold of Ciba-Geigy. Burpee's became the largest vegetable and flower seed operation in North America. Then, in the space of little more than a decade, it passed through General Foods and ITT to Wickes.[15] Vilmorin-Andrieux succumbed to the advances of Limagrain in the early 1970s.

Both in the nineteenth century and the early part of the twentieth, farmers banded together in cooperatives for purposes ranging from collective bargaining for fair prices to plant breeding and seed distribution. In Sweden, Swalof, one of the country's two dominant breeding establishments, is the child of both government grants and the Swedish farm cooperative movement. Its seeds are now sold as far away as Nicaragua. In Germany and Austria, Baywa and branches of the Raiffeissen cooperative movement play a key role in getting seed to farmers. Many of the major seed enterprises in France are formed into cooperatives as well, including Limagrain. Limagrain also shunned its humble beginnings to spread its wings across Europe and the ocean to North America—as did the two dominant Dutch breeders, Cebeco-Handelsraad and Suiker Unie.

In the United States, the Land O' Lakes and FFR cooperatives have enlarged their repertoire to include breeding. Each has had to make its alliances, however; Land O' Lakes recently made a deal with the Farm Bureau in the U.S. for the marketing of its seeds and chemicals. Across the border in Canada, cooperatives such as Manitoba Pool Elevators and the Saskatchewan Wheat Pool share a limited interest in plant breeding but continue to be the largest distributors of farm seed.

For city folk who nurse an image of the farm co-op as a rustic manifestation of prairie populism, where heavy-booted farmers hack out strategies to stave off starvation, the offices of today's big co-ops would come as something of a shock. If only the topsoil were as deep as the pile of those boardroom carpets! While vestiges of the old populism and purpose still occasionally surface on ceremonial occasions such as annual members' meetings, most of the old co-ops have been co-opted by the multinationals they were bred to fight.

According to Austria's Hill Farmers, a well-organized band of innovative farmers clinging to Alpine slopes and remote valleys, three multinationals—Pioneer Hi-Bred, Ciba-Geigy, and Continental Grain—have used the co-ops as the market entry vehicle for their branded seed. Much to the chagrin of the co-ops' plant breeding affiliates, the three companies evidently approached the retail side of the movement with deals that assured the retailers a higher profit margin on imported seed than they were able to win from their own varieties. The companies seemed willing to accept short-term losses (or lower profits) in order to introduce themselves to Austrian farmers and to commandeer Austria's major farm supply network. Once strong and independent national breeding programs are now being subverted to the menial task of testing overseas varieties and multiplying seed.

France's Jean-Pierre Berlan, an economist with Institut Nationale por la Reschereche Agronomie (INRA) (the leading government agency charged with agricultural research), noted that multinationals moved into his country the same way: first offering high profit margins to establish their brands through the often effective co-op distribution system, and then shunting the co-ops aside and taking over directly once their own footing was firm.

For some, the term "cooperative" borders on fiction. France's Limagrain and Holland's Suiker Unie (despite its eighteen thousand beet growers) behave much more like multinationals than farmer-based organizations. In effect, the farmers have as much to say about their "co-op" as most shareholders have to say about the running of Ciba-Geigy or Occidental Petroleum—two other seed companies. Since ministries of agriculture around the world place great store in the co-op movement and often base proposals for new legislation on the benefits that will accrue to co-ops, it is important to separate romanticism from reality. Without besmirching the goals of co-ops or the crucial role played by many small and effective co-ops in many countries, many others are nothing more than combinations of companies.

Many of the world's new seed companies have come by their position logically enough. For cereal seed, the means of production are also the end product for consumption. It was almost inevitable that flour millers and grain traders would become involved in the industry. It was hardly surprising, then, that Britain's giant miller, Ranks Hovis McDougall, should also rise to become that country's biggest cereal seed seller (before selling its seed interests to Dalgety-Spillars in 1984). Likewise, the spread of Cargill, Continental Grain, and Archer-Daniels-Midland into seeds caused hardly the batting of an eyelid on the Chicago Board of Trade.

The arrival on the scene of livestock feed suppliers like Bibby's (now owned by Barlow-Rand of South Africa) and Grain Processors Corporation can also be regarded as only natural. When firms such as Pillsbury, Campbell Soup, and Central Soya of the United States (which sold its seed interests to Upjohn in 1983) and Unilever, Inchcape and Sinclair McGill of Britain also acquired seed houses and/or moved into plant breeding, brokers were only marginally more surprised.

MULTINATIONAL ALCHEMISTS

But the world of the old-time seed company was jarred irreparably when the hounds breathing hard down the acquisition trail were seen to be transnational petrochemical and pharmaceutical concerns. Our research shows that since 1970, multinational giants ranging from Shell Oil to ITT have moved to buy or otherwise control nearly a thousand once independent seed companies. Because of the almost total absence of benchmark studies, the exact breakdown between acquired seed houses and those that have been developed by larger firms cannot easily be determined. The real number of takeovers and the total holdings of transnational enterprises are likely much greater than can be tabulated from annual reports and trade publications.

Large firms have moved rapidly to dominate the seed business in many countries. In Britain three firms—two of them foreign (Ciba-Geigy of

Switzerland and Volvo of Sweden)—control nearly eighty percent of the garden seed market. In the Netherlands three companies control seventy percent of the agricultural seed market and four companies—only one Dutch—have ninety percent of the market for horticultural seeds. One firm has tied up thirty-eight percent of the lucrative U.S. corn seed market.

After years of careful study, partially funded by the Dutch government, Ton Groosman, Anita Linnemann, and Holke Wierema of the Tilburg Development Research Institute estimate that transnational seed houses control about a quarter of seed sales in the North and between five and ten percent of sales in the expanding Third World market. They also calculate that the industry in the South is worth about $3.8 billion per year.[16] Table 3 offers three interpretations of the top ten seed companies in the world today.

Eleven of the dominant thirty companies are part of the chemicals complex. Only six are traditional seed houses. Even of these six, KWS has surrendered large share holdings to Volvo of Sweden and Hoechst of Germany, and the two top Japanese seed companies appear to be closely allied to Japanese petrochemical interests.

Corporate concentration has bred catalogue concentration as well. As the size of the companies controlling the traditional seed trade has grown, the size of their truly distinct offerings has shrunk. The aggressive sale of new varieties has also sped up the pace of genetic erosion as older, heirloom varieties are bumped off the market or out of the fields by glossier newcomers. Both farmers and gardeners are affected. In 1857, the proprietors of Wethersfield Seed Gardens listed twenty different types of turnips. In contrast, the current catalogue for Burpee's—a seed company that has passed through the digestive tracts of two transnationals in fifteen years—offers today's gardener a choice of only two varieties.

Dr. James Thomas, formerly with Agriculture Canada (the Canadian equivalent of the USDA) and now with the Consortium for International Development (a collaborative effort of six U.S. universities), notes that "Bolivia . . . is already feeling the effects of reduced 'source of seed' competition because the multinationals have already purchased a number of smaller suppliers and then only provide one 'most profitable' line from among those purchased."[17] As indicated in chapter four, this commercial emphasis not only limits the number of varieties made available, but also causes the disappearance of traditional types. It is widely recognized that Germany's largest seed company—Kleinwanzlebener Saatzucht (KWS)

Table 3
Top Ten Global Seed Houses Rankings by Three Sources

Economist '87	Ag. Gen. Report '88	Teweles '89
Pioneer	Pioneer	Pioneer
Shell	Sandoz	Sandoz
Sandoz	Dekalb-Pfizer	Upjohn
Dekalb-Pfizer	Upjohn	Limagrain
Upjohn	Limagrain	Cargill
Limagrain	ICI	Volvo
ICI	Ciba-Geigy	ICI
Ciba-Geigy	Shell	France Mais
LaFarge-Coppee	LaFarge-Coppee	Dekalb-Pfizer
Cargill	Cargill	Claus

Sources: Economist, 15 August 1987, p.56
Agricultural Genetics Report, Sept/Oct 1988, p.2
Seed Industry, May 1989, p.13

single-handedly mopped up the gene pool for beets in Turkey with its introduction of the Detroit Globe strain some years ago.[18] Similarly, an American eggplant—Black Beauty—has succeeded in wiping away the immense diversity of eggplants that once populated the Sudan.

Writing in the newsletter of the International Board for Plant Genetic Resources, Dr. Brian Ford-Lloyd and Dr. Peter Crisp contend that considerable genetic diversity can still be found in the stocks of small European seed companies, but that this situation is changing because "small companies are rapidly being taken over by larger, often international companies which are more interested in supplying varieties acceptable over a wide area, with the result that local types aie superseded."[19]

As severe as the consequences of amalgamation of catalogues may be for genetic erosion in the garden, the wider marketing strategies of international companies have much more extensive effects. Prompted in considerable measure by the opportunity for exclusive monopoly patents, the new seed houses have driven many varieties and crops into local oblivion.

Predictably, the role of transnational enterprises in genetic erosion is much disputed by the companies themselves, as well as by some scientists and governments. One of the most passionate corporate champions has been a crusty octogenarian named Sir Otto Frankel, a scientist long involved in genetic conservation. In speeches and papers delivered in Canberra, Vancouver, and Rome, Frankel has characterized our contentions that companies and their patents contribute to genetic erosion as unfounded, unscientific polemics.

But it was also Frankel—on the subject of livestock breeding—who decried the "commercial—hence biological—monopolies which stifle genetic diversity."[20] He went on to quote his colleague, J.M. Rendel, who suggested that society "limit by legal measures the range of a person or company producing breeding stock." Thus, added Frankel, "ensuring that local differentiation is maintained, say, at the level of a state in terms of the U.S.A. . . ."

Divested of its niceties, Rendel's (and, by implication, Frankel's) proposal could restrict the spread of any new variety to an area no greater than Texas or the Ukraine. Multinational genetics supply companies would either have to do some work or throw in the towel.

But this is not the first time the issue of the monopolization of germplasm and the connection between corporate monopoly and conservation have been made. Nikolai Vavilov declared his opposition to genetic monopolies several decades ago when he complained about the virtual monopoly in livestock breeding held by the British in the nineteenth century.

And the National Academy of Sciences 1972 study, *Genetic Vulnerability of Major Crops,* concluded that "this [genetic] uniformity derives from powerful economic and legislative forces."[21] One of the authors of that study was Peter Day, then with the Connecticut Agricultural Experimental Station and later director of the Cambridge Plant Breeding Institute. Invited to address the Committee on Genetic Experimentation of the International Scientific Union in 1980 on the causes of uniformity, Day cited pressure to use high-yielding varieties, the tendency to rely on clonal propagation methods, and (reading from committee minutes), "the involvement in plant breeding of agribusiness aggravated by plant variety rights legislation . . ."

Following an agonizing internal debate, the FAO secretariat also linked the patenting of seeds to the related issue of genetic uniformity. In a 1980 interdepartmental memo on FAO's relations with the Union for the Protection of New Varieties of Plants (UPOV), the plant patenting agency in Geneva, FAO could hardly have been more critical of the negative effect of patent monopolies: "The subsequent commercial competitiveness resulting from . . . UPOV has led to intensive breeding of new varieties on a limited genetic base, resulting on several occasions in widespread disease epidemics."[22]

MALCHEMY

An intense international debate regarding the chemical industry and its involvement in and impact on plant breeding has arisen in recent years. By 1979, corporate sensitivity to the charge that chemical firms were moving into seeds was such that the industry was hotly denying the invasion—arguing that only the occasional firm had actually been acquired and that the overall impact was insignificant.

That same year, however, the respected D.J. van der Have Company of Holland offered its hundredth anniversary study of trends in the industry, making a point of the recent acquisition activity of chemical companies.[23] The study noted that the reasons for the takeovers were "not all clear" and warned against the dangers of corporate monopolies. Van der Have was itself later acquired by Suiker Unie, to become the centerpiece in that company's array of nine seed houses in five countries.

The jig was up when, in 1981, Thomas Urban of Pioneer Hi-Bred told Ann Crittendon of the *New York Times* that chemical companies were indeed moving into plant breeding. The reason? "The assumption behind the trend," Urban said, "is that the new owners can improve the plant's resistance to the herbicides and pesticides that the parent company sells."[24]

The chemical industry has closer links to seeds than might at first be recognized on Wall Street. Most of today's leading establishments have their origins in the fibers and dyestuffs trade. Such origins are invariably linked to plants. The scramble to monopolize plant-based dyes such as indigo, and the discovery in 1891 of a process to produce rayon from the mulberry bush, combined to launch some of Europe's dominant firms.

The pharmaceutical industry is doubly connected to seeds. Apothecaries from the earliest times worked with herbs to formulate both magical and very real cures for human maladies. Even now at least a quarter of our prescription drugs are based upon plant material, and most of the remaining drugs are synthesized from plants.

The new alchemists' second connection to seeds is more immediate. It was a Bayer drug researcher in 1935 who discovered that a red dye used in the textile industry was also useful against pneumonia and scarlet fever. The race was on and companies poured every liquid concoction known and unknown to humankind into test vials in the hope of curing something.

The synthesizing of chemicals from plants opened the door to the petroleum industry. "Petrochemicals" became the great feedstock upon which plastics, resins, pharmaceuticals and pesticides were forced to depend. Atlantic-Richfield, Occidental Petroleum, Standard Oil of California (Chevron), British Petroleum and Royal Dutch/Shell all moved heavily into chemicals to become raw materials suppliers to the pharmaceutical trade, and to undertake their own work in producing agricultural chemicals.

It was only a matter of time before the oil giants would discover seeds and vertically integrate into plant breeding. Of these giants, none looms so large over agriculture as Royal Dutch/Shell—by its own admission, the oil industry's biggest pesticides manufacturer, and now one of the largest seed companies in the world.

Inheritance taxes facing the heirs of family seed companies and the sudden over-liquidity gushing in the petroleum trade in the 1970s created the preconditions for some takeovers. According to takeover tycoons, companies such as General Foods and ITT became interested in the seed trade due to a desire to "move closer to the consumer" in times of economic change.[25] Others gave credit to the lower bank borrowing rates available at the beginning of the seventies (in the United States) and the growth of investor confidence.

Merck offered a reason for the drug companies' move into seeds and fresh produce. Increasing government intervention on behalf of consumers was lowering drug prices at the same time as stringent safety regulations were creating havoc with the research and development (R & D) cost for new drugs. Linked to this was the growing complexity of drug research following the "easy" victories of the 1950s and 1960s. In short, pharmaceutical companies were finding profits harder to come by.

Similar forces were threatening the profitability of pesticide manufacturers. Health and environmental concerns were crimping corporate style and the freewheeling R & D of the preceding decades appeared forever gone. Worried executives could stare down the road and see a long-term reduction in the use of crop chemicals. Diversification into plant breeding and seed sales seemed an especially attractive hedge since seeds and pesticides required a similar marketing system, and since any decline in use of chemicals would likely necessitate sowing more seed and a greater reliance upon new plant varieties.

The R & D directors of corporate drug and pesticide divisions could feel comfortable with seeds. All three products take upwards of six to twelve years to come to market; all three require extensive testing of thousands of chemical compounds or breeding lines; all three areas are "science-based" and need heavy capital outlays.

By 1970, there was another connection. With the passage of the U.S. Plant Variety Protection Act at Christmas, seeds joined pesticides and pharmaceuticals in offering control over new products through patent-like protection. Without the kind of monopoly control over marketing and licensing afforded by patents, chemical companies would have been without one of their most important weapons.

Different chemical companies found themselves in the seed business for different reasons. But whatever the reasons that induced family firms to sell out may have been, a historic base in botanical sciences and the similarities in R & D requirements played a major role in causing chemical firms to jump so dramatically into plant breeding. Whether this was premeditated or not, it is hardly surprising that once companies entered the seed business, their marketing departments took note of the opportunity to bring together seeds and agricultural chemicals.

Some chemical companies are even institutionalizing the link between seeds and chemicals. When Reichold Chemicals bought out Florida Seed and Feed a few years ago, the company's annual report disclosed that the lawn and garden concern would be organizationally linked to Reichold's two crop chemical interests (Woolfolk and Sunniland), which were being moved to take advantage of the Florida possibilities.

Upjohn, the new owner of Asgrow Seed and O's Gold, has consolidated agricultural R & D in a new seven-storey complex at Kalamazoo, Michigan. In a single complex, Upjohn is developing plant varieties, plant growth regulators, pesticides, animal growth regulators, and animal genetics programs. And across the Atlantic, Volvo's Miln Marsters cereal breeding subsidiary has a new research station at Docking in England where industry sources claim Volvo's major cereal breeding and chemical testing will be concentrated.

In the 1940s it was this nascent connection between seeds and chemicals that began a process that would create the green revolution. Despite theories to the contrary, the fertilizer-seed partnership begun with the green revolution was much less the outgrowth of a cunning plot by fertilizer manufacturers than the inevitable consequence of the extension of corporate western agricultural models to the Third World. Trained at land grant colleges closely allied to farm input companies, green revolution scientists carried their biases happily into Asia, Africa and Latin America. No plot was necessary. As a young man, Norman Borlaug left the biochemical laboratories of E. I. duPont de Nemours and Company to take up wheat breeding at the Rockefeller Foundation-supported research center in Mexico. This connection was to give him a Nobel Prize for Peace for breeding the "miracle strains" of wheat that gave force to the green revolution. Borlaug's early training in plant disease problems and his history with DuPont certainly influenced his attitude toward crop chemicals in his breeding work.

The varieties developed by Borlaug and his counterparts at other research stations around the world caused yields of crops like wheat, maize and rice to skyrocket—but only under certain conditions. In order to sustain yields, the new varieties needed fertilizers and irrigation. Skilled at translating more nutrients into more grain,[26] the high-response seeds encouraged the use of more pesticides because the fertilizers also promoted weed growth, and irrigation stimulated insect development. In turn, the increased use of chemicals killed off fish in the rice paddies and wiped away a major source of income and protein for Asian farmers.[27] Meanwhile, the genetic uniformity of the crops added pressure for the employment of yet more insecticides and machinery.

The green revolution failed to live up to its promise of solving the problem of world hunger. It failed because the problem was not simply one of too little food and thus could not be solved simply by producing more. The problem was and is one of maldistribution, and ultimately lack of power and opportunity amongst the hungry in Third World countries to participate in the process of food production—and consumption. Unfortunately, by offering tantalizing yields and profits to the handful of Third World farmers able to invest in the new seeds and the required inputs, the green revolution helped further concentrate rural wealth and power in the hands of a few—exacerbating the very process that had helped create so much hunger in the first place, and the very problem so many had claimed the green revolution would solve.

But the green revolution did unquestionably accomplish one thing: it opened up the world to agrichemical corporations. The new seeds were dependent on fertilizers, pesticides and farm machinery. And the dependency provided new markets for Shell Oil, Ciba-Geigy, Monsanto, Massey-Ferguson, and a number of other giant transnational corporations. In the space of a few decades the face of world agriculture changed dramatically at the hands of these corporations. As Hilkka Pietila, of Finland's (farmer-oriented) Centre Party, told an audience in Helsinki, the last century of scientific breeding has divorced farmers from their own seed and effectively reduced the number of breeders from millions of farmers to a few hundred seed companies.[28]

Peasant agriculture succumbed. Peasant crop varieties vanished. And companies that originally got into the game to supply crop chemicals found that they could make even more money selling varieties dependent on those chemicals. Greater yields were good for humanity, the companies told the public. But dependency on greater yields was even better for business.

In its early years, the green revolution concentrated global research on just three crops—wheat, corn and rice. Few Third World countries had any substantial research infrastructure at the time. Nevertheless, the green revolution altered the agendas of the research efforts that were in place. Scientists were encouraged to focus on the three crops. Innovative work on integrated pest management in India[29] and research on millets and sorghums in Africa, for example, were abandoned.[30] By the late 1970s, political pressures had pushed the IARCs into broader work on farm management and research into almost thirty crops. Some sources estimate that no more than one-third of the total IARC budget now goes to wheat, rice and corn—leaving the remaining twenty-seven to squabble over the rest of the budget.[31]

Not surprisingly, concerns are widely expressed about the marketing strategies of companies capable of offering their customers both seeds and crop chemicals. Such package deals go back to at least the early twentieth century. German sugar beet farmers have often found themselves forced to accept a seed/chemicals combination through their contracts with sugar refineries. A young German farmer at an agricultural meeting in Wurtsberg outlined one such story for us. Sugar beet growers work under contract with a regional refinery. Usually there is not much choice. The refinery sends around a list of recommended seed varieties. That year, there was only one recommended monogerm variety offered by Volvo. Their premier variety could only be bought "pelleted" in a clay and chemicals package intended to provide seed size uniformity and a head start in the ground. Farmers had no choice but to accept the Volvo package.

The *pièce de résistance* in the package business is a triple cocktail now being marketed by Ciba-Geigy. According to Dr. M.H. Hahmoud of the Sudan Mechanized Farming Corporation, Ciba-Geigy has attempted to sell the Sudan government its own brand of hybrid grain sorghum packaged with three new, patented Ciba-Geigy chemicals.[32] Two of the chemicals, Bicep and Milocep, are intended to control a wide range of weeds. The third, Concep, is known as a herbicide antidote especially created to protect the plant from Ciba-Geigy's patented Dual herbicide that might otherwise damage the seed. As industry journals reported, this package has the "advantage" that farmers must return to the company each year to have the chemicals injected into the seed coating. The net effect of all this creative chemistry is that Ciba-Geigy can expand its sales of the Dual herbicide while wrapping up the Sudan's sorghum crop in a package very profitable for the company. The Sudan Mechanized Farming Corporation rejected the deal. Undaunted, Ciba-Geigy was back in the Horn of Africa—in Ethiopia this time—in the early summer of 1985 offering to sell hybrid sorghum to the famine-ridden nation.[33]

Similar motivations may have influenced Sweden's Kemanobel. The chemical giant has long been Europe's self-confessed leader in fruit and vegetable fumigants. Now it is also Europe's leading retail packet seed supplier. Its garden seeds—often wrapped in chemicals—dominated the trade in continental Europe in the early 1980s.

Stauffer Chemicals (first bought by Cheeseborough-Pond and then carved up and sold piecemeal to Unilever and ICI) followed suit when it scooped up three U.S. seed companies in the late seventies to create Stauffer Seeds. All three constituent companies had been engaged in the corn seed trade and Stauffer just happened to have a leading edge in corn herbicides. Now, along with Stauffer corn hybrids, farmers can buy their Sultan and Eradican herbicides and Dyfonate soil insecticide—all from the same company. Likewise, Ciba-Geigy, a dominant force in the corn broadleaf herbicide market, acquired Funk Seeds—one of the largest corn seed breeders in the world.

Some pharmaceutical/chemical enterprises like Upjohn had no real history in crop chemicals when they began buying into the genetics supply business in the seventies. Nevertheless, Upjohn, for example, took over TUCO Chemicals and established "crop protection" research operations in Florida and Michigan. Now a major breeder of soybeans and field peas and beans, Upjohn, in 1982, was also able to release its first in-house agrichemicals—for beans.

When opponents to the chemical company takeovers in the seed industry first raised these concerns in the late seventies, industry spokespeople attempted to laugh them out of the debating halls. Granted, chemicals and seeds research could be dovetailed, but no one from Ciba-Geigy to the Agricultural Institute of Canada was prepared to concede that such arduous R & D would be worthwhile. Dr. Bryan Harvey, chairing a Canadian Agricultural Institute team that looked into this question, observed that "plant breeding is hard enough now without adding on new problems such as a chemical connection."[34] Perhaps so, but today members of that same institute are hard at work breeding a canola (nutritious form of rapeseed developed in Canada) variety that will tolerate the herbicide Atrazine, originally patented by Ciba-Geigy.

Nevertheless, many companies insist that the chemical and seed aspects of their work are kept intentionally at arm's length, arguing that it is not feasible to dovetail chemical and seed R & D. But other companies contend that this very connection can benefit farmers, and that any artificial division of R & D must be regarded as bad business.

Dalgety-Spillars, a major breeding force particularly in peas and beans in the U.K., now offers a "pea wheel" which farmers can dial for advice on all aspects of dried pea production—fertilizers and pesticides included.

Mike Carver of Shell's Nickerson Seeds subsidiary claims that British farmers can save up to twenty pounds (sterling) an acre by following the input guide booklets Shell distributes free with each of its varieties. The booklets detail the need, timing, quantity, and application procedures of growth regulators, fertilizers, and pesticides for each variety. Competitors' chemicals are included in the booklets but some are worried that Shell will have a bias toward its own proprietary products, if for no other reason than that it knows them best. Could it be otherwise? Shell has recently developed a seed treatment known as Panoram 25, intended to combat loose smut in spring barley. Shell's spring barley varieties occupy over half the U.K. market. It is hardly by chance that the company has focused on seed treatments for this crop, and it is absurd to think Shell would ignore a pesticide marketing strategy that affected its own seed varieties.

It is Shell Chemicals, after all, that patents the parent firm's seeds in Italy and also in South Africa. Shell Petroleum handles company seeds in the

West Indies and, in the U.S., Shell Development Corporation is working on sterility-inducing chemicals for its hybrid wheat program. Shell features both its corn varieties and its herbicides in the same advertisements in German magazines.[35] And why not? The chairman of Shell's seventy or so seed subsidiaries was the head of Shell Chemicals and, as an advertisement in African business journals proudly proclaims, "We Grow Them and We Protect Them Too."[36]

It may well be that modern breeding methods virtually guarantee the chemical connection. Some tough talk on the dangers of "chemical breeding" arose from an FAO gathering of experts on integrated pest control in 1978. Dr. O.M.B. de Ponti of Holland's Wageningen Agricultural University pointed a finger at the world's crop variety testing plots. In an unpublished, uncompromising paper critical of the practices of public and private breeders alike, De Ponti argued that the use of insecticides and herbicides on test plots would obscure the pest resistant characteristics of the varieties being tested, and would inevitably lead to the decline and loss of the genes that could confer protection.[37] The very method of developing varieties was guaranteeing the world's long-term dependence upon chemicals.

The key here is that chemical companies have more opportunities to profit from plant breeding than do traditional family seed companies. The exotic work now being undertaken in seed coatings and plant growth regulants undoubtedly benefits large corporations more than family companies.

Historically, chemical corporations have profited from cartel and other price-fixing agreements that have effectively eliminated competition.[38] This was clearly shown in 1960 when the EEC demanded that all industries disclose any accords that would violate the new antitrust provisions in the Treaty of Rome, which created the Common Market in 1957. The agreements unveiled by the chemical industry were second in number only to those for metals, and included a host of price-fixing and territorial arrangements.[39] Observers at the time estimated that only a fraction—between five and fifty percent—of all the cartel agreements were actually made public. In the years since, many, if not most, of the companies now owning or building seed houses have been the targets of antitrust and combines investigations on both sides of the Atlantic.[40]

As the same group of companies becomes dominant in the seed industry, it is reasonable to expect that some of the same commercial practices will prevail. With territory and licensing agreements, chemical companies *could* become freer to neglect disease resistant plant breeding without fear of competition.

Indications of this neglect are already visible. In a 1979 study by the U.S. Office of Technology Assessment—an agency of the U.S. Congress—officials noted a trend toward the use of borer-susceptible hybrid corn varieties, resulting in increased infestation and greater use of insecticides.[41] Likewise, the report pointed to a marked decline at the time in the breeding of wheat varieties resistant to wheat stem sawfly and Hessian fly. The report added that "commercial seed companies have also de-emphasized efforts to incorporate insect and even some disease and nematode resistance into new cultivars." Criticism was leveled at the public research stations that had withdrawn from this breeding work on the assumption that private firms were taking over. This trend, which can be traced to the advent of new U.S. seed patenting laws in 1970, is estimated to have cost farmers in one Great Plains state more than 25 million dollars in 1978 alone.[42]

A similar failure in disease resistant plant breeding has been reported in the United Kingdom. On the top floor of a rambling old building housing the National Institute of Agricultural Botany (NIAB) in Cambridge, Valerie Silvey produced a study in 1978 that may still be rattling the windows of Cambridge today.[43] Silvey painstakingly traced the evolution of British wheat and barley varieties from 1947 to 1977 and attempted to attribute yield increases proportionately to improved husbandry and variety improvement. The British seed trade and its counterparts as far away as Australia have used the study to "prove" that commercial plant breeding is a smashing success, for according to Silvey, wheat yields in the United Kingdom have grown steadily throughout the thirty-year period.

But barley yields have not. In the sixties and seventies—during the time in which barley became a major U.K. crop—variety improvement dropped well behind that of wheat. Silvey credits the failure to a "single-gene resistance" approach in plant breeding. Varieties being offered in the sixties and seventies contained only one gene, often the same gene regardless of the variety, that was intended to ward off a pest or disease.

The breeding strategy led to what Silvey describes as a "boom and bust" cycle. Farmers would grab up the hot new variety that offered virtually complete protection from the pest, only to find that the gene had been overcome by mutating pests within three years. The result, says Silvey, has been a rapid turnover of new varieties and an increase in the use of fungicides.

Down on the ground floor of the NIAB, Dr. J.K. Doodsen, the institute's energetic pathologist, claims that the now rejected single-gene strategy still has currency among some commercial breeders today (Doodsen refers to them as "chemical breeders").[44] According to the pathologist, some companies rely much too heavily upon crop chemicals to protect new varieties. Doodsen points to the case of a Volvo barley variety known as Golden Promise, which was a smash hit in Scotland until its susceptibility to mildew allowed pathogens to build up in such strength as to overcome even more mildew resistant varieties. As a consequence, Scotland now has a serious mildew problem for the first time ever—and the problem is moving south into England. Golden Promise is heavily treated with fungicides, and fungicides must now also be used on once resistant barley varieties.

The British problem is acute enough to have precipitated debate within the agribusiness community itself. Peter Wormell, a longtime farm reporter and author, penned a heretical article in *AgriTrade-News* in the spring of 1981, asking "Are High Yields Necessary?"[45] Wormell asserted that the new varieties were "even more prone to catch the common cold than some of their more robust ancestors." Referring to the country's historic problems with rust, the journalist added sadly, "Today there is a chemical to do the job."

In short, the future may offer chemical/seed conglomerates a choice of how they want to make their profits: through pesticides, or through disease and pest resistant plant breeding. This likelihood was inadvertently identified by the Economic Commission for Europe in a 1982 study of the chemicals industry.[46] The ECE noted that breeders had three objectives: yield; machine harvesting and processing; and disease resistance. It added that breeders were happy to "sacrifice" disease control where pesticides were available. If the breeder is also the pesticides manufacturer, that sacrifice hardly hurts at all.

But farmers are far from stupid. Given a choice between a disease-resistant plant variety and one that depends on chemicals, they will opt for the low-input variety. Until the widespread decline in public breeding began, a genuine choice was almost assured. Public varieties were released only when they were thought to be superior to existing varieties. As University of Wisconsin sociologist Jack Kloppenburg notes, "Public breeders disciplined the market as to quality, price and structure . . ." thus curtailing the opportunities for abuse.[47] Farmers could reject the chemical connection. Now that the dominant seed providers are chemical concerns, however, the options are dwindling.

The chemical breeders have been with us for less than two decades and it is still far too early to assess the implications of their involvement. It takes at least eight years to bring a new pesticide to market, and as long or longer to develop a new crop variety. The real impact of the linking of these two major agricultural inputs will not be clear until the 1990s.

The point is this: chemical companies can profit in two different ways when a disease attacks a crop. They have a choice, in fact. They can either find an "organic" plant breeders' solution to the problem by developing an improved variety, or they can simply refer farmers to a chemical already on the shelf. No bad faith on the part of breeders is required. Chemical firms need only trust in the biological laws that lead pests and diseases to mutate and ultimately overcome the defenses of new varieties. Family seed firms have no such advantages. First, they lack the crop/chemical profit incentive, and secondly, unlike the multinationals, they lack the economic clout to protect their markets from other seed companies.

Perhaps the most noteworthy accomplishment attributable to the new breed of seed companies in the past dozen years has been the vanquishing of its public sector competition. Until the 1970s, much of the innovative force in plant breeding came from universities and state-run institutes. The giants of plant breeding were INRA in France, the Commonwealth Scientific and Industrial Research Organization (CSIRO) in Australia and the Plant Breeding Institute (PBI, now Unilever) in Britain. These large scientific centers had the talent and discipline necessary to engage in long-term breeding strategies that paid off every year with corn in France, new forages in Australia, and Maris Huntsman wheat and Proctor barley in the United Kingdom. Now these public institutions—the last barrier preventing complete domination of plant breeding by multinational petrochemical corporations—are dying.

Speaking at an Andean seeds symposium in Lima, Peru, in the late seventies, Dirk Boringer—a high seeds official with the Bonn government—talked of breeding work shifting out of public hands to the private sector in his country. Today, seed researchers in German universities largely devote themselves to basic research in germplasm and other aspects of agronomy. This research is then made available to companies at rates subsidized by German taxpayers. At Hohenheimer University in

Stuttgart, cereal breeding still takes place, but the breeders are constrained by a committee of companies to make the final varieties available to the trade!

In 1983, then U.S. Secretary of Agriculture John Block announced that "federal research is being phased out of conventional plant breeding programs where the private sector can meet these needs."[48] Seed companies have no interest in seeing government institutes compete with them in plant breeding. Company officials prefer to talk in terms of a "division of labor" in which public bodies develop new breeding material which is turned over to companies for "final exploitation in the marketplace."[49] Translated, this means that government does the costly, basic and innovative research, while big companies pick up the profits in the marketplace. In the U.S., government will soon not be competing with companies in variety release, and Department of Agriculture authorities now recommend that scientists engaged in breeding be moved to "germplasm enhancement" projects.[50] The companies have won.

Will the public sector continue to release new varieties in the nineties? Yes. We will see government-funded research leading to new varieties for a long time to come. But will the new releases simply struggle to meet the needs of farmers sowing minor crops or farming in "fringe" microclimates (both unattractive to multinationals because of the limited demand)? Or will government breeders be forced to abandon the high cost of supporting these markets as well? Will public research be able to compete with the R & D budgets and advertising dollars of the big corporations involved in the breeding of the major crops? Whichever way the public sector goes, its ability to influence and regulate the private sector is rapidly becoming history.

Ironically, the private sector will continue contracting with university professors to further the companies' research needs more directly, as pointed out by Martin Kenney in his book, *Biotechnology: The University-Industrial Complex*.[51] But the decrease in public support for university plant breeding will surely detract from the ability of the universities to train the future plant breeders that will be needed by the private sector.

The United States is not alone in its retreat. In 1970, when the august Plant Breeding Institute at Cambridge threw open the doors of a new complex of laboratories, scientists there were looking forward to a boom in agricultural research. Then, on the eve of PBI's seventy-fifth anniversary, the Thatcher Government auctioned off the world's premier public breeder to Unilever—the highest bidder among contenders rumored to include ICI, Shell and Ciba-Geigy. On August 6, 1987, Unilever bought PBI for sixty-six million pounds and entered the ranks of the world's largest seed companies.

The seed industry has a long history. But its real roots lie with today's subsistence farmers and their and our Neolithic ancestors. Long before Darwin, Mendel, or the renowned American plant breeder Luther Burbank, the plant breeding facet of agriculture was undertaken by sophisticated, capable people who walked their fields with a keen eye for the best plants to be saved for seed. As Vavilov more than once noted, crop landraces were the result of intelligent, innovative minds—and often the work of geniuses.

Today the number of seed innovators has declined, as have the numbers of truly different varieties available, and the genes that make them up. This has not come about as the result of corporate plots, nor is it the fault of science or scientists. But the science has been channeled to an elite rather than shared among those who grow our food. In the process there were gains, but much was lost.

About the only legacy of many an old family seed company today is its name, now followed by the phrase "subsidiary of . . ." It is not nostalgia that causes concern over the takeover of the seed business by multinational petrochemical and drug corporations. On the eve of the spread of new biotechnologies, control of the seed industry is a necessary prerequisite to the control of all agricultural markets. In order to ensure their control, the new genetics supply industry managers are also looking to patent seeds, genes, and the very processes of life.

CHAPTER SEVEN

Enter Biotechnology

We've invented fire. The sky's the limit.

—Waclaw Szybalski,

University of Wisconsin

Pork chops growing on trees. Cows the size of elephants. If company press releases and media reports are correct, these are the sorts of startling inventions biotechnology could produce in the future.

Biotechnology is a term used to describe a variety of techniques which involve the use and manipulation of living organisms to make commercial products. Tissue culturing, cloning, and genetic engineering are among the best known techniques of biotechnology. With biotechnology, scientists are increasingly able to move genes coding for a certain characteristic from one organism to another. And they are finding it easier and easier to produce countless identical copies of a plant, for example, by culturing and cloning cells from the original plant.

Though hailed as one of the greatest scientific breakthroughs of all time, biotechnology does not yet rank alongside the Neolithic revolution which gave us cultivated crops. Indeed, despite the intimidating terminology, some aspects of biotechnology are mundane. Virtually everyone, from the gardener who digs up and multiplies strawberry plants to the amateur beer brewer, has had some experience with biotechnology without knowing it. The difference is that scientists are speeding up such processes, making them more sophisticated, and applying them more broadly.

Most significantly, they are learning how to transfer genes between species. Genes from fireflies have been transferred into tobacco plants.[1] And at the University of Kentucky, scientists have spliced a gene from a flounder (a salt water fish) into a soybean plant in an attempt to make soybeans more tolerant of cold weather.[2]

The advent of biotechnology raises many questions about the role of technology in our society and about the possible economic, social and political impacts of such a powerful new tool. It is not the purpose of this book to explore these issues. (For a discussion of them see "The Laws of Life," in the journal *Development Dialogue*, 1988:1&2, written by the staff of the Rural Advancement Fund International.) This chapter will briefly and simply focus on the ways in which biotechnology might affect genetic diversity.

LIKELY EFFECTS

Biotechnology is certainly life in the fast lane. Various techniques now being developed are allowing plant breeders to cut drastically the time required to produce new varieties. At SunGene of San Jose, California, a computer scans two million "plants" in a petri dish in a matter of hours, looking for the useful variations it would have taken Luther Burbank acres of land and months to find.[3] Breeding, testing, evaluation, and multiplication are performed with dispatch with the aid of genetic mapping, genetic engineering and tissue culture. For our immediate purposes it is not important to delve into definitions of these techniques. It is important to realize what they are doing to the process of plant breeding and varietal development. In annual crops like soybeans and wheat, the time needed to produce a new variety can be halved. For perennial crops like oil palm, breeding time can be reduced from thirty or forty years down to just seven or eight.[4]

The pace quickens after breeding. At Plant Genetics, Inc., of Davis, California, scientists take a single gram of callus (a cluster of undifferentiated plant cells raised in the lab) and "brew" as many as ten million plant embryos in a cultivation tank in six months. Then, in quantities of twenty thousand to a batch, the embryos are encapsulated, enveloped if so desired in herbicides, fungicides and fertilizers, and sent off to the market as "artificial seeds."[5] Rather than plan seed production and warehousing logistics at least two years in advance and risk the uncertainties and expenses of seed multiplication conditions, the genetics supply industry is positioning itself to brew whatever embryos it judges marketable scant months ahead of the sales orders, circumventing growers and dealers. The encapsulated embryo will be the conduit for all the agricultural inputs at once. And by providing for disease-free plant propagation, the new methods will drastically reduce disease transfer problems in seed shipments, thus facilitating exports.

What characteristics will the new varieties have? All sorts of promises are being made. At the least we can expect significant yield increases. Some observers are predicting that yields in developing countries should be at least forty percent higher than today's [1988]. And they further predict twenty to forty percent increases in developed countries.[6]

In individual crops, yield increases could zoom higher. In oil palm, tomorrow's premier vegetable oil, production could skyrocket over two hundred percent as a result of both varietal improvement and the rapid multiplication and dissemination of superior oil palm clones.[7] Intensive cropping systems combined with new varieties of cacao developed through biotechnology could result in increases of a stunning 750 percent over today's average.[8]

The impact of such yield increases on the genetic resources—both traditional landraces and the modern varieties now in the field—could well be devastating. In chapter four we discussed in detail the process whereby traditional landraces are being replaced and chased into extinction by modern varieties. Farmers historically willing or forced to abandon traditional varieties for modern ones promising modest yield increases—typically less than five percent a year[9]—will find they have no choice but to sow their fields with the products of biotechnology. In the competitive agricultural marketplace, farmers who decline to take two hundred percent yield increases will be looking for new jobs.

The pace of biotechnological breakthroughs is so fast that one could safely say that no genetic conservation system exists which could collect the traditional varieties as quickly as they will likely be eliminated by biotechnology.

CUSTOMIZED CROPS

Research at the Rural Advancement Fund International indicates that at least sixty-five research programs worldwide now focus on breeding herbicide tolerance into agricultural crops. Corporations

like Monsanto, DuPont, Ciba-Geigy and Bayer are w
genes that will impart resistance to certain herbicides—
brands—in order to sell seeds as part of a package
increase chemical sales. The ecological danger lies in the
often bordered and invaded by weeds which are related
field. This is particularly so in the Third World, but i
the U.S. with crops as disparate as sorghum, radishes, lettuce, and sunflowers.[10] As we learned in chapter two, crops "introgress" or cross with their wild relatives. Recent studies at the University of California indicate that "gene flow by pollen occurs readily among populations that are separated by 100 m, 1,000 m, and even greater distances."[11]

Dr. David Ehrenfeld of Rutgers University says "it will only be a few growing seasons before we can expect to see this engineered herbicide resistance transferred naturally, in the field, to the weeds themselves."[12] This could result in the need for more and stronger herbicides, particularly in the Third World. And it could have unforeseen consequences for the delicate and complex coevolution that takes place amongst plants in field and forest out of our sight. The scientists at California, obviously concerned that engineered genes other than those conferring herbicide resistance might escape to other plants, conclude that "without substantial mitigation . . . ecological damage seems likely, or even certain."[13]

And there are further complications. Companies such as Unilever and DNA Plant Technology are developing new crops like thaumatin, a West African rain forest shrub with a berry a hundred thousand times sweeter than sugar.[14] Other companies are working on methods to produce the flavoring, vanilla, in the laboratory, eliminating the need for both the vanilla orchid and the vanilla farmer. If commercialized, such developments could wipe out crop production and eliminate a wealth of genetic diversity in the replaced crops precisely at a time when it would be most difficult to argue that the diversity should be saved.

But, of course, it must be saved. We have no way of knowing how the genetically narrow-based monocultures of biotechnology will stand up to the rigors of the real world. Genetic engineers can design and engineer to their hearts' content. But they will always be limited by incomplete understanding of the infinite interactions between plants and the environment. Surprises are already taking place. Oil palm clones planted in 1983 began producing abnormal flowers and fruits in 1986. As a senior Unilever scientist reflected, "Field tests on clones produced in the lab went well, but when we went from lab to scale-up, problems occurred."[15]

To the extent that it is successful in producing attractive new varieties, biotechnology will displace existing varieties. The more startling the breakthrough, the more rapid and complete the extinction of traditional landraces and cultivars will be. But the unheralded changes will also take their toll. The ever narrowing crop genetic base that will be created by biotechnology, protestations notwithstanding, will create added problems. We will have to pray that the new miracle crops and animals produced by biotechnology will work in the field, and continue to work, because they will go a long way towards eliminating all other options.

IRONY OF BIOTECHNOLOGY

Change is the constant in our world. New pests and diseases are always emerging. The climate is changing, perhaps rapidly. Industrial society is altering everything from our eating habits to the chemical composition of the soil and the rain that falls upon it. Expanding human populations are pushing agriculture onto new, often marginal, lands. All of these developments place a burden on genetic resources to provide the necessary adaptations.

While the potential of biotechnology is great, delivery has been slow and sometimes disappointing. Genetic mapping, a method for identifying and facilitating the use of genetic variation, has not proceeded as fast as some predicted. Genetic engineering without the knowledge of where genes conferring certain characteristics are located on the chromosome is robbed of much of its potential.

Furthermore, the complexity of interactions of genes may complicate the transfer of genes across species lines. As we noted earlier, researchers have successfully inserted a flounder gene into the cell structure of soybeans, but that gene has yet to express itself and scientists cannot explain why. Even genetic engineers need models. Finding these models in the already existing plants of even the most remote areas of Amazonia may prove easier than inventing them from scratch in the form of "artificial genes." Were our knowledge of the environment complete and our techniques foolproof, genetic engineer Julian Davies might be right in implying that all the genetic diversity we need could be found in the "parking lot."[16] But we are not there yet and we may never be.

The genetic diversity now in existence represents a wealth of already adapted and tested genetic combinations. Even with the impressive tools of biotechnology, it will take years to assess, let alone understand, the diversity we already have available. We have scarcely begun to exploit this diversity or even survey it. Our gene banks, as inadequate as they are, are underutilized by plant breeders.[17] Yet scientists now talk of designing genes to order and creating new species to meet every imaginable need.

Others, including ourselves, have spoken of genetic resources as the "raw materials" for plant breeding and biotechnology. But they are not raw materials in the same sense as iron ore or copper. In the ground, these materials have not been improved. Even after extraction, no human genius or act of inventiveness is evident in a shovelful of ore. The landraces of peasant farmers represent *improved* materials. They embody the thoughts, insights, inventiveness and hard work of farmers past and present.

Farmers may not wear white coats or use fancy equipment, but they observe variation, note mutations, practice selection, and engage in breeding and seed multiplication—the basic activities of their plant breeder counterparts in the North. The difference is that in industrialized countries, laws are designed to recognize the achievements of individualized work. The "inventor" gets a patent or patent-like protection. The Third World system of innovation is more informal and communal in structure, hardly conducive to our formalized patent system. Thus, the contributions of farmers go unrecognized, unrewarded, unprotected—even denigrated.

Ironically, the need of industrialized countries to rewrite their patent laws in the wake of biotechnology may afford the Third World an opportunity to push for recognition of the informal innovation system. As discussions in various United Nations fora explore the need to encourage and reward those engaged in biotechnology, the possibility of establishing a system of rewards for the Third World arises.

The diversity in hand, deemed by some to be irrelevant, should be seen as a hedge against the possible ill effects of biotechnology—as insurance against mistakes or unrealized potential. It is troubling that the infant science of biotechnology, as promoted by the public relations departments of its corporate sponsors, might lull us into thinking that we have a reduced need for diversity, thus lessening our commitment to conservation.

How ironic that this might be happening at the same time that biotechnology fashions its wonders from the raw material of genetic diversity. And how odd that the assumption that evolution is completely controllable might cause scientists themselves to devalue and perhaps overlook the diversity and variability upon which evolution depends—and upon which their own careers are being built. And how naive we would have to be to think that impoverished Third World nations would eagerly donate their raw materials to the biotechnology industry in exchange for the fabled pot at the end of the rainbow. The poor have rarely trusted the rich completely.

Concern over the loss of genetic diversity at the hands of agricultural biotechnology should not be interpreted as revealing Luddite tendencies in the authors. The technology is exciting and has great potential in areas beyond the scope of this book, like health care. But, like technologies that have preceded it, biotechnology will most likely come with unintended and unforeseen consequences—among them the displacement and extinction of traditional crop varieties.

Diversity gives us options. It is best to keep our options open. At the advent of the age of biotechnology, the question "Why save diversity?" can best be answered by another question: What diversity can we afford to lose?

CHAPTER EIGHT

Global Conservation Begins

All species are potential Humpty Dumpties: the processes of evolution, as we know them, will not put them together again on this planet once they are destroyed.—David Ehrenfeld

---

Government realization that the world's genetic diversity needs protection has come slowly, beginning with the early work of Vavilov in the 1920s and fragile initiatives at the United Nations in the late forties, and building toward proposals for a UN Food and Agriculture Organization (FAO) Commission on Biological Diversity in the late eighties.

The road to realization has been rocky. Collection efforts launched in the 1960s have been diverted to a technical international body not answerable to governments: IBPGR, profiled later in this chapter. In the campaign to conserve, the short-term interests of the North have dominated, leaving most of the South's collected germplasm in cold storage in the North. Those same short-term interests have also meant poor funding for long-term storage and a battery of numbing technical conundrums that have caused genetic erosion *within the gene banks* to become a major problem.

When Nikolai Vavilov first met his staff at the All Union Institute of Applied Botany and New Crops in 1923, he warned them that their task was not only to gather plants for the immediate breeding needs of Soviet agriculture, but also to save seeds from extinction. The great geobotanist and ultimate plant explorer was the first to recognize genetic erosion as a global threat to food security, and to see seed keeping as a conservation strategy.

Few heard his warning outside the Soviet Union. One who did was Harry Harlan, this century's other inveterate plant explorer and an old ally of Vavilov's. As mentioned, Harlan first warned in 1936 that genetic erosion was taking place due to the introduction of new crop varieties.

Plant collectors, whether working for current breeding programs or dedicated to the long haul, need a place to store their seed safe from pests, dampness, and heat. The Russians claim collections dating back well into the last century and it is probable that every agricultural research institute going back at least to the 1870s had modest working collections. Vavilov may also have been the first to think of seed storage to guard against extinction. Certainly, the Russians were at the forefront of this work. The Leningrad gene bank dates back to the 1920s. Both the Germans and the Italians were collecting barley in Ethiopia in the 1930s (Harry Harlan was there in 1923 and Vavilov scoured the country in 1927) and the seed is understood to have been placed in "long"-term storage. The Austrians have had a seed bank high in a Tyrol glacier since the late 1930s and in the United States—although a National Seed Storage Laboratory was not established until 1958—some short-term facilities have existed at least since World War II.

By the 1960s, as the green revolution took hold, few if any governments had even a paper strategy for the conservation of genetic diversity. With rare exceptions, the work being done was sub-national, university collecting bereft of national financial support. This limited action was restricted to crop or even problem-specific seed collecting and storage in the plant equivalent of meat lockers—cold rooms with a modicum of temperature and humidity controls—intended to maintain the viability of seeds for several years.

In the 1970s, as the green revolution reached gale force and was blowing furiously across the fields of Asia, Africa and Latin America, only a thin disconnected line of badly-funded, makeshift gene banks and a handful of dedicated and isolated plant collectors stood in its way. Vavilov was dead and Harry Harlan had passed on the torch to his son Jack. Here and there, in obscure and esoteric journals, eloquent calls were issued. Almost no one heard.

The history of gene banks is not long. Short histories encourage shortsightedness and tend to focus on individuals and away from the background movements and events that push people onto the world's stage. Thus, it is easy to describe the history of seed conservation by talking about Vavilov and the Harlans, Bennett, Frankel, Swaminathan, and T.T. Chang. The history can also be told as a pillar-to-post succession of UN conferences and international conventions.

While the people and the places are important signposts, the underlying reality in the early 1970s was that the green revolution was wiping out genetic diversity even as the uniformity in the Soviet wheat crop and the U.S. corn crop was demonstrating the political and economic costs of genetic vulnerability. New advances in microbiology and new corporate opportunities in the seed industry, combined with the ecology movement that built up after the sixties, made germplasm conservation a social, political, economic, and corporate necessity.

But there were people involved as well. The history is short enough for most of the major actors still to be active.

Erna Bennett's detractors cite the FAO founding conference in Quebec City in 1946 as evidence that the intergovernmental work on genetic erosion did not begin that early spring day in 1966 when the Irish breeder first strode into FAO's Circus Massimus headquarters. In a sense, they are right. A decade after Harlan's 1930s warning—and two decades before Bennett came to FAO—the opening session of FAO's Committee on Agriculture did discuss genetic resources. And FAO did, with the International Biological Programme (IBP), host a key conference on plant genetics back in 1961.

But it was the colorful, outspoken Ulster-born Irish revolutionary who first coined the phrase "genetic conservation" and brought substance and strategy to the term for the world community.

Recognizing how little was actually being done, Bennett spent her first year with FAO not in Rome but at the Plant Introduction Center at Izmir in Turkey, where she collected and selected seeds that launched conservation work in the Near East center of diversity identified by Vavilov. In 1967, she returned to Rome to organize the second FAO/IBP Conference on Genetic Resources, and engineered the resolutions and political momentum necessary to create a Crop Ecology Unit in FAO the following year.

Denied the directorship of the unit (because of being female and because she had already acquired notoriety in the FAO hierarchy), Bennett joined forces with Harero León of Costa Rica to fashion a modest secretariat out of the funds allocated via FAO's Plant Production and Protection Division.

Among her accomplishments are the creation of a documentation center and the distribution of the first worldwide newsletter drawing attention to the crisis of genetic erosion. Perhaps more important—and despite intense opposition within FAO—Bennett insisted that the organization get out and collect itself, before it was too late.

In the late sixties and early seventies, her angular figure could be seen from the edge of the western Sahara to the highlands of Afghanistan, surveying, gathering, and reporting. Between sometimes dangerous expeditions (she was variously threatened by bandits, nearly killed on mountain roads and, on one occasion, thrown into a harem on the assumption that she was the subordinate to her male driver), Bennett edited the first classic work on genetic resources with Sir Otto Frankel. Published in 1970, this book brought together all the wisdom available at the time and forced the scientific community to take notice.

Based on the book and Bennett's newsletter evangelism, Otto Frankel—the cranky Austrian-born, Australia-based scientist—was able to persuade the 1972 Stockholm Conference on the Human Environment to adopt a resolution calling for concerted global action on the conservation of crop genetic resources.

The Stockholm meeting brought the issue of genetic conservation—everybody's orphan—perilously close to popularity. It also set in motion a battle over the control of the crop genetic resources "mandate" in the international community.

Among those vying for control was the Consultative Group on International Agricultural Research, better known as CGIAR or "Cigar". A year before the Stockholm gathering, the Rockefeller and Ford Foundations had cajoled industrialized countries into capitalizing on the glory of the green revolution by putting together a whole new agricultural research network that could sidestep the politics and bureaucracy of the UN system. The new creature was CGIAR and under its umbrella, once privately-funded centers like CIMMYT and IRRI were able to make common cause for greater funding. It is arguable that this additional money might otherwise have accrued to FAO. Others maintain that the UN agency was uninterested in research and would never have been in the running.

More significantly, foundation-backed crop research institutes including CIMMYT and IRRI were now considered "IARCs" (international agricultural research centers) affiliated to CGIAR. Suffering from political overexposure in the Third World, the institutes hoped to find solace and security with pseudo-UN protection. CGIAR set up housekeeping at the World Bank headquarters in Washington. This pseudo-UN status was all-important. The foundations could give the appearance of moving the IARCs into the UN fold while at the same time creating a donor-driven forum which virtually excluded normal North-South political realities—the best of all possible worlds.

In 1971, however, genetic resources were not seen as part of the CGIAR mandate. The push was on the rapid expansion of green revolution technologies and the creation of IARCs in new regions to deal with more crops.

Only after the founding meetings did it become evident to the CGIAR hierarchy that donor states might have money available for germplasm conservation. The southern corn leaf blight of 1970 in the United States had softened at least one monied donor into believing that work on genetic conservation was urgent.

CGIAR made its move early in 1972. Otto Frankel, as Erna Bennett's non-UN sidekick, was invited to Beltsville, Maryland, to join a working party instructed to come up with a global "strategy." Bennett—among the few in the world with proven practical as well as theoretical knowledge—was deliberately excluded. While Frankel and others might have hoped otherwise, the objective of the exercise was to prove that not enough was being done and to establish CGIAR's ascendancy as the "expert" body capable of tackling the problem. Not fully attuned to organizational nuances, Frankel penned a multi-year plan proposing a network of regional gene banks. But his idea of a network did not provide for centralized control by donors.

T.T. Chang of IRRI arrived at the Holiday Inn on Baltimore Avenue on Sunday, March 20, looking for a soft bed after a flight that covered half the globe.[1] No such luck. Otto Frankel was already on hand and pressing hard for his master plan. Chang recalls the meeting as arduous and frustrating and, years later, would describe it as the "infamous" Beltsville meeting. It was a heavyweight crowd. Jack Harlan and John Creech represented the United States, Jack Hawkes flew in from the U.K. and Dieter Bommer came from the German Gene Bank at Braunschweig. T.T. Chang and Mario Gutíerrez of CIMMYT represented the IARCs while Jorge León was the sole FAO defender. In true UN style, a Costa Rican, a Turk, an Indian and an Ethiopian were drawn in for local color. Chang had to do double duty as the only Chinese person present.

The surprise participants were two scientists from the Vavilov Institute in Leningrad. With the winter wheat crop doomed in the Ukraine, the Soviet Union was definitely interested in genetic resources. Chang recalls that the senior Soviet scientist went out of his way to be friendly until he learned that Chang's passport was from Taiwan.

Despite the rivalries, Otto Frankel prevailed and a proposal came forward for a highly decentralized network of nine regional gene banks. The budget was in the neighborhood of three million dollars.

But CGIAR did not favor decentralism. Its Technical Advisory Committee (TAC), which had convened the Beltsville meeting, rejected the plan two weeks later at a session in Rome. Frankel, Hawkes and Chang met in Izmir, Turkey, later in April to redraft. But CGIAR was still unenthused. The whole effort was finally turned over to M.S. Swaminathan, a non-Beltsville participant, drastically revised to create space for a central germplasm "traffic cop," and ultimately approved in the heat of midsummer Washington a year later. Ten months after TAC's assent, the International Board for Plant Genetic Resources (IBPGR) held its first session in Rome. By then, T.T. Chang was back in the Philippines tending his rice collection. And Jack Harlan was trying to sound the alarm, writing articles with titles like "Genetics of Disaster."

Always a bit of a loose cannon, Frankel refused to take CGIAR's evasions for action and stormed the gates of the Stockholm environment conference that summer to press his case before the world's environmentalists. Heavily focused on whales and pandas that year, most environmentalists couldn't see the drama in crop seeds and did nothing more than pass a sympathetic resolution. But it was all just enough to open the way for yet another new creation—the UN Environment Program (UNEP)—to spring up and claim a piece of the inaction.

By the end of 1972 (with the Beltsville report still in the political limbo of a "draft"), CGIAR in Washington and the fledgling UNEP were both eyeing genetic resources as a part of their natural mandate. Meanwhile, back at the FAO, Erna Bennett was using all her considerable talent to rescue wheat in drought-ridden Afghanistan.

Somewhere between Beltsville, Izmir, Rome, Stockholm, and the final tabling of the Beltsville report in Washington, the IBPGR was created when the "mistakes" of the Beltsville plan were "corrected" by the eclectic Swaminathan, then scientific advisor to the Indian Cabinet.[2]

In the end, IBPGR was made a member of the CGIAR network but assigned offices and staff within FAO. UNEP—weakest of the three combatants—was thrown the bone of a permanent seat at the IBPGR table.

A key player in the final decisions was Richard Demuth, a Washington lawyer attached to the State Department, with a heavy involvement in agriculture in the Third World. Demuth worked with Swaminathan in the development of IBPGR and became its first chair. Retiring from the State Department in 1974, Demuth re-established his law practice and also took on the position chairing CGIAR's newest star, IBPGR. Thus IBPGR, self-described as a purely technical, scientific body, chose a Washington lawyer with State Department connections as its first chair.

In the Byzantine political environment of the United Nations, Demuth was—and is, despite radical heart surgery in his mid-seventies—a major wheeler-dealer. (Otto Frankel might have seemed a likely candidate for the chair—or at least the board—but his political naivete nearly cost CGIAR its prize and did nothing to endear him to FAO.)

The rise of IBPGR to a view overlooking the Circus Massimus in 1974 was something of a puzzle, most of all to the little crew surrounding Erna Bennett and the Crop Ecology and Genetic Resources Unit high on the seventh floor of FAO's Building C. IBPGR was not the Beltsville dream of nine omnibus regional seed banks necklaced by a ring of crop-specific banks and a bankroll big enough to choke off genetic erosion. It was—rumor had it—a "catalytic" coordinating body intended to stimulate collection and conservation, and to build a gene bank network from existing institutions.

The funds for this work were largely to come from FAO. The staff, and this was the chief source of bemusement to the folks on the seventh floor, were to be borrowed from FAO's Crop Ecology Unit. In summary, the good people at CGIAR and UNEP were responding to FAO's call for action with a takeover of FAO's action, funds, and staff to create a little structural diversity and sow the seeds for future adversity. Bennett and FAO were the victims of an interagency mugging.

What was happening between the Crop Ecology Unit and IBPGR must be understood in the context of the politics of the early seventies. Norman Borlaug's Nobel Prize in 1970 had convinced the industrialized donor ure of agricultural development lay in green revolution he same time, the evolution of the debate between North e United Nations system had contributed to developed bout working in the UN. New structures such as CGIAR N politics and perhaps address some of the real and imag- ined about the quality of work at the FAO.

The industrialized countries and their scientists could thus "save" the Third World without the frustration of accountability. Genetic resources—as the building block for green revolution varieties—were to be the preserve of CGIAR and its international agricultural research centers. The fact that the work was already under way in FAO—and had a long history there—was an inconvenience overcome with a little juggling.

More than half a century after Vavilov had sounded the alarm, the international scientific community could claim to have the beginnings of a global seeds strategy.

THE SYSTEM?

On the eve of its tenth anniversary celebrations, the International Board for Plant Genetic Resources was to report an impressive list of achievements. More than three hundred collection expeditions had been mounted, yielding close to 120,000 new seed accessions covering 120 species in eighty countries.[3] The work of collecting and conserving germplasm had been conducted in cooperation with more than 550 "catalyzed" scientists and an almost equal number of institutions. More than 700 scientists underwent IBPGR-supported training programs and IBPGR produced more than 300 publications, ranging from crop descriptor lists (for documentation of gene bank material) to catalogues of crop collections and major texts on seed storage, quarantine, and rejuvenation. The heart of all this activity was said to be a "network" of global and regional base collections for thirty-eight crops housed in major gene banks under written agreements to assure their long-term survival and free exchange—all this on a pitiful annual budget that only broke the five million-dollar barrier in 1985.[4]

These wonders were accomplished by a modest coordinating staff of six scientists at headquarters and two in Washington, supplemented by five regional officers based in Syria, Thailand, Burkina Faso, Kenya, and Colombia.[5] At times the staff was considerably smaller.

IBPGR's work is nevertheless only part of the global conservation effort. The agency for technical cooperation in the Federal Republic of Germany, Deutsche Gesellschaft für Technische Zusammenarbeit (GTZ) has helped to establish long-term storage facilities at Addis Ababa, Ethiopia, and Turrialba, Costa Rica. The Inter-American Development Bank has assisted in Brazil. New storage facilities were provided in Thailand and banks have been or are being established in Sri Lanka (with support from Japan), Pakistan (World Bank), India (United Kingdom), China (Rockefeller Foundation), Bangladesh (Asian Development Bank), Peru and Chile (with Japanese support), and Bulgaria (UN Development Bank). The Nordic Gene Bank, a joint Scandinavian venture, is working with the nine Southern African Development Cooperation Committee (SADCC) states to develop a major gene bank for southern Africa at Lusaka in Zambia, offering more than thirty million dollars over a twenty-year period to ensure long-term stability. At the same time, Germany's GTZ is developing a gene bank in Kenya. Other conservation supporters include FAO, the UN Environment Program, the Ford Foundation, and the government of Australia.[6]

CGIAR officials claim that the total global commitment to plant genetic resources conservation was $60 million annually by the mid-1980s.[7] Pinning the figures down is not easy. About $10 million comes via the CGIAR-affiliated IARCs, and then there is the IBPGR's $5 million. Another $13 million comes from one country—the United States. It would require an act of faith to believe that these sums are actually being used for the collection and conservation of genetic resources, however, as will be seen below.

On the outskirts of the "official" system is the International Union for the Conservation of Nature and Natural Resources (IUCN) and its sister body, the Worldwide Fund for Nature (formerly the World Wildlife Fund) which, in 1984, moved on from pandas to plants. While governments have concentrated almost exclusively on the collection and storage of endangered seeds in gene banks (so-called "*ex situ* conservation"), IUCN has focused on wild plant species and the need to establish biosphere reserves ("*in situ*").

With the lone and humble exception of IUCN, the international scientific community has defined its conservation strategy in terms of crop-specific collection expeditions with the gathered seeds placed in cold storage in temperature and humidity controlled gene banks. Worldwide, FAO estimates that at least half a million seed samples are now in storage.

Interestingly, IBPGR's crop data books indicate closer to two million stored samples.

Erna Bennett—who never much cared for CGIAR's mugging of her Crop Ecology Unit, and whose vocal criticisms of IBPGR won her an early-retirement offer—finds these collection figures more awful than awesome.[8] More to the point, Third World delegates to FAO conferences over the past several years have grown increasingly critical of IBPGR's structural ambiguities and self-serving statistics.

While criticism is to be expected—and is, in our opinion, justified—it must be recognized that the "system" devised by individual governments and institutes and later by IBPGR came about in an economic (if not political and intellectual) vacuum, in a situation that can accurately be described as an emergency. Much of what has been done has been done the fastest, surest way possible. Some of the niceties of international diplomacy and some of the finer aspects of genetics have been sacrificed. This, in a sense, is to be expected.

Nevertheless, the conservation campaign has amounted to a drive to collect a few select eggs and place them in a single, tattered basket. Lawrence Hills of Henry Doubleday Research Association in England points out that this is like saving a "representative sample" of Goyas and Rembrandts.

Aside from the technical critique, Third World governments are also concerned about who is storing the Goyas and Rembrandts. For a quasi-UN body, IBPGR can be judged either incredibly gauche or unabashedly political in having opted to base an overwhelming share of germplasm samples in western industrialized countries, and in the United States in particular. Both the technical and the political critiques deserve closer examination.

So successful has IBPGR been, according to IBPGR, that it ended its first decade with radical plans to revamp its mandate, and move away from its original collection emphasis toward germplasm documentation and evaluation. According to its own data, most of the genetic diversity of key cereal crops is now safely banked. Most of the gene banks needed to preserve diversity are already "onstream."

This optimism appears to be founded upon two reports, one by the Rockefeller Foundation circulated by IBPGR in early 1985, and the other released by IBPGR through a CGIAR review panel about the same time.[9]

According to IBPGR's own review, 95% of the crop genetic diversity of wheat, potatoes and barley and samples of 90% of the world's corn are now safely locked away in gene banks. A little less, 80% of sorghum and 70% of cultivated rice and groundnuts, are also safe. Among less famous crops, okra is also 80% secured and cowpeas weigh in at 70%. The scene is not quite so rosy for sweet potatoes (60%) and the board concedes that beans, at a lowly 20%, need more attention. Nevertheless, the percentages—after only a decade—are nothing short of spectacular.

Without fanfare, IBPGR and the Rockefeller Foundation were actually telling the world that they had a head count on Creation, or at least on Creation-collected. Linnaeus would have been amazed. While scientists had never before ventured even to guess the total number of higher-order plant species to the nearest fifty thousand, IBPGR was bravely announcing that farmers and nature had produced (more or less) 110,000 wheat variants (including landraces) and between 12,000 and 12,500 wild wheat types, and that almost the whole kit and caboodle were safely tucked away in gene banks.

A closer examination of the two studies, however, reveals problems. For example, although the studies concur on the quantity of germplasm in storage, they differ drastically on how much of that material is "unique." For the nine crops the two studies share in common, the gap is about 125,000 samples. Further, although both studies presume to know the total genetic variability of their nine crops—and their percentage estimates of the amount of variability in safe storage are not very different—their actual head count differs by 180,000 accessions. Although some of the differences may be modest, some are enormous. Using the Rockefeller Foundation's figures, the total diversity of corn would amount to about 41,000 accessions. IBPGR (through CGIAR) suggests 67,000. For potatoes, the foundation's count is about 35,000 but this would meet only slightly more than half of IBPGR's estimate that the total diversity of the potato crop is closer to 63,000 varieties. All in all, the IBPGR and the foundation have a one-third or greater difference of opinion.[10]

While most people in the Third World remain largely unaware of or uninterested in genetic resources, Third World ambassadors at FAO, at least, find the discrepancies disturbing. IBPGR has used the figures mentioned to justify a shift away from collecting and into breeder-oriented germplasm evaluation work. Representatives of developing countries are not convinced that the record of plant collecting even for the major cereal crops warrants any sense of security. They are also concerned that progress on the dominant cereals front is now being used as an excuse to move on and neglect regional and poor people's crops.

Seed collecting could be described as a scientific adventure that uncovers an art. In the early days there was often more adventure and artistry than science. Plant explorers would leap from their burros to shake seed from a novel looking plant and then trundle on randomly in search of more exotica. In time, this knight-errant approach gave way to a more sophisticated "conspicuous consumption" technique, wherein collectors would window-shop through a farmer's field in search of any visible variation. While this was an improvement, the diligence was heavily biased toward plant morphology and did little to preserve more subtle characteristics. Such "seat-of-the-pants" strategies are probably responsible for most of the germplasm in gene banks today. A less subjective grid-sampling approach to field collecting has been encouraged by FAO and IBPGR, but it has a short history.

Most seed collectors suffer from a kind of botanic xenophobia that makes them tiptoe through the millets, stomp through the teff, and utterly disregard the medicinal herbs on their way to a much needed barley.[11] All might be endangered, but only the barley has commercial value back home. In fact, an international expedition may be so mission-oriented and driven by the specific need to solve a disease problem that members will search exclusively for crop varieties hinting resistance to such a disease. The rest is left in the fields uncollected and unreported. Future expeditions can easily be misled by pins on a map to believe that a region has already been scientifically collected.

Even today, plant explorers find it hard to venture far from their Land Rovers. Expedition leaders tend to be Europeans with a definite fondness for their creature comforts. Even the more vigorous and rigorous often collect under time pressures calling them back to their universities and test plots.

Erna Bennett, puzzled by the pattern of cereal collections in Turkey, discovered its logic when she overlaid the collection flags with a road map. Virtually without exception, seeds had been gathered along the sides of major thoroughfares. (Cynics may now argue that the salvation of seeds might best be furthered by new highways and extension of hotel franchise . . .)

Not that the grid system is the only way to gather diversity. Indomitable plant explorers Brian Ford-Lloyd and Peter Crisp let their fingers do the walking through the pages of the Rome and Naples phone directories. By telephoning thirty family-owned seedhouses, they uncovered a hundred previously uncollected samples of cauliflower, broccoli and red beet that had escaped the notice of the Italian gene bank.[12]

Regardless of the collection procedure, the reporting has left much to be desired. According to IBPGR's own evaluation, sixty-five percent of all samples lack even rudimentary passport information—making it very difficult for any scientist to conclude that the nameless sacks in the storeroom account for any precise percentage of the crop's diversity.

"How much" and "where" are major concerns—but so is "what." Until 1980, the overwhelming share (three-quarters or more) of all seeds collected by IBPGR was for the top cereal crops. After 1980, the scope expanded to include legumes and potatoes, but very little else.

While an emphasis on key cereals is logical, these are also the crops of primary interest to northern breeders. Endangered crops of major regional importance—crops which have limited economic value but are valuable sources of nutrition for poor people—have been ignored.

The issue of poor people's crops—including medicinal plants—grew to the point of disrupting a technical meeting we attended that was convened by IBPGR in conjunction with FAO and UNEP in the spring of 1981. In particular, Latin American states were incensed that IBPGR was determined to keep its focus on high-value international crops. At one point, the session almost broke down completely as Third World scientists felt themselves ignored. When debate moved from the technical to the political level during the FAO biennial conference later that same year, IBPGR was moved to be more cooperative.

To be fair, IBPGR had begun to diversify its collections by 1980, expanding into food legumes, forages and root crops. By the launch of its second decade in 1984, the board was proposing to reduce its emphasis on cereal crops from 56% of all samples in the first decade to 25% of collections. Food legumes would decline from 20% to 15% but other crops—fruits, vegetables, root crops, and woody species—would jump from 24% to 60% of collections.

All this was hardly a great victory for the people of the South, however, since IBPGR was not only slicing the pie differently, but also turning the pie into a tart. From spending close to a third of its total budget on collections in the first decade, the board was proposing in its annual report to reduce collection funding to one-fifth of future budgets. Thus, although the proportion of funding for fruit, vegetable, and minor species collection would expand sharply, the actual numbers of plants to be gathered would hardly change at all.

Then in the mid-eighties, IBPGR suddenly changed horses in mid gene pool. Without explanation, the organization went from warning the scientific world that genetic erosion was catastrophic to claiming that the crisis was very near to being solved. Behind this switch was Trevor Williams, executive secretary of IBPGR for most of its short history, and eventually director. Williams had told a 1980 Union for the Protection of New Varieties of Plants (UPOV) symposium: "Every time we receive expert advice the warning bell is ringing in the more remote parts of the world. Genetic erosion is advancing at a rapid rate."[13] It was also Williams who warned at a technical conference in 1981 that "all too often . . . sampling has been done only along major roads and that the main interest was useful looking plants rather than representative genetic variability."[14] In 1982, he advised readers of *Nature* magazine, "If the work is not done in the next five to ten years, we're finished."[15]

Yet when Williams wrote the introduction to IBPGR's 1983 annual report, he felt secure in announcing a number of major policy changes: "Specifically, the widespread collection of cultivars will be slowed down except for documented emergency situations . . ."[16]

What had happened during 1983 to change IBPGR's stance? Some ten thousand seed samples were collected during the year. As good as this was, it was well below the previous year's count and a far cry from the halcyon days of 1978 when almost eighteen thousand accessions had been recorded. Was the chasm between calamity and security so readily bridged? Or were the warning bells merely muted? Dr. Gary Nabhan of the Desert Botanical Garden in Arizona offers another possible explanation: the North's battered old banks were awash in seeds that had not yet been evaluated, and word went out from IBPGR to slow collections in order to give curators a chance to catch up and grow out their accessions. But IBPGR has never confirmed this slightly less embarrassing explanation.

Whatever or how much IBPGR has collected does not bear any automatic relationship to what and how much has been conserved. Although lip service is paid to conservation strategies such as biosphere reserves (a kind of national parks scheme for wild plants) and botanical gardens, IBPGR and the international community have placed virtually all their genetic eggs in the same basket: the gene bank.

LIFE IN A GENE BANK

Stripped to its inelegant basics, a gene bank is a seed store—an insulated room housing collected seed samples in small containers at low temperatures and controlled humidity. Samples are often held in tin cans but may also be found in glass jars, laminated aluminum foil envelopes, or plain old paper bags. The idea is to keep the seed clean, cold, and dry, and to disturb it as little as possible for as long as possible. Under ideal conditions, wheat seed may remain viable for 390 years and some barley types can reportedly spring back into life after 33,500 years.[17] In so-called long-term gene banks, samples are stored at −10 to −20 degrees centigrade for several decades, or even a century. In medium-term facilities, seeds are maintained at zero to five degrees centigrade for up to twenty years, and in short-term banks, seeds may be held at no better than room temperature and may only survive a few years.

Seed of every crop and every variety of every crop will respond to life in a gene bank differently. No matter how tough the seed, a sample will eventually degenerate to a point where it must be rejuvenated (grown out in a greenhouse or field, harvested, and put back into the gene bank). On average, a gene bank should probably grow out a tenth of its contents every year, or recycle its whole collection every decade.

Gene banks may not save everything but they can do a pretty fair job. The issue is not that the gene bank "basket" is a bad choice for the eggs that will keep the great-grandchildren fed in the year 2100, but that banks are IBPGR's only basket. Other choices could be added.

In November of 1979, the first volleys of the international debate over the control of plant genetic resources had just been fired. M.S. Swaminathan was taking advantage of his position chairing FAO to introduce the issue to the biennial session of the FAO Conference. IBPGR was undergoing its five-year review by CGIAR. And K.E. Prasada Rao was climbing into a four-wheel drive Toyota Land Cruiser in Khartoum, along with three Sudanese colleagues, in order to recapture sorghum material "lost during maintenance."[18]

Before the month was over, Rao had scoured the Gazira provinces and the Blue Nile region of the eastern Sudan in search of threatened landraces and wild sorghums. Hampered by a chronic shortage of fuel, the Toyota could not take them everywhere, but Rao and company talked with farmers, searched through markets and prowled the banks of the Blue Nile in an increasingly desperate search for two strains—Hagira and Zera-zera—which had made a major contribution to U.S. sorghum breeding early in the twentieth century, and genes from which form the basis for all modern hybrid grain sorghums today.[19]

Hagira was not to be found and only fifteen samples of Zera-zera were finally uncovered near Damazin, "on the verge of extinction."[20] Although well planned, the expedition suffered all the customary problems of long-range collecting. Time pressure dictated that the team stick close to the Land Cruiser, and that meant keeping to the passable roads. Somewhere over the horizon, the last stands of Hagira may have been waiting.

There were other problems. Since the trip was set to coincide with the sorghum harvest, the team was bound to miss the season for pearl millet, which was their crop of second priority. Rao had to content himself with unrepresentative samples from village markets. In addition, sorghum strains like Kurgis from the hilly areas around Kurnuk could not be harvested until the month after Rao was back at the International Center for Research in Semi-Arid Tropics (ICRISAT) in India.

Six years later, in July, 1985, we caught an early morning flight from the Delhi airport for Hyderabad and ICRISAT, a CGIAR system IARC with responsibility for sorghum and millets. Rao had been on the ICRISAT payroll when he collected in the Sudan in 1979 and we understood that the samples were stored at the institute's gene bank since the Sudan cannot afford a gene bank of its own.

After a long and tortuous ride among cows and cars, we saw ICRISAT looming on the parched summer horizon as an impressive complex of low modern buildings. Within were vine-covered walkways, ancient Tamil artifacts and modern microcomputers. Aside from a giant water tower, the most imposing physical feature was in the central courtyard—a large stone urn, used a millennium ago to store grain and seed for the next year's crop. If you follow the walkway past the urn far enough, you will eventually come to the genetic resources unit with its offices and cold rooms.

Our hope was to visit Dr. Melak Mengesha, the highly regarded Ethiopian leader of the ICRISAT Genetic Resources Unit. Regrettably, Mengesha was on sabbatical in Iowa at the time of our visit and his stand-in—the retired head of the Ames, Iowa, gene bank, Dr. Willis Skrdla—while strong on corn, knows sorghum only by reputation. During a brief tour, we walked through the seed processing warehouse and asked about the procedure for drying. Cereal seed must be dried slowly and at low temperatures so as not to endanger its long-term viability in the bank. Dr. Skrdla admitted that seed was customarily dried under the hot Indian sun—that day bordering on forty degrees centigrade. The IBPGR recommended temperature is fifteen degrees.

When we entered the first cold storage unit, it took our guide and us a few moments to realize that something was wrong. We were warm. Shirt-sleeved repairmen had replaced the usual warmly-cloaked seed librarians and were variously tinkering with pipes and mopping up puddles that covered the floor of sorghum's most important artificial gene pool. In the embarrassment, it was difficult to ask how long the unit had been malfunctioning, although even inexperienced observers could see that such a heavily insulated room does not turn balmy overnight.

Not that Zera-zera sorghum is gone forever, or that the whole ICRISAT collection was destroyed. The institute has several storage units and only one was damaged. Most—but not all—of the seed in that unit probably survived the shock of changed temperature and humidity. Our concern is that ICRISAT's sorghum experience is not an isolated incident and shows clearly how problems of collection can be compounded by problems of conservation when all your eggs are in the one proverbial basket.

Recent history has shown that gene banks are as prone to failure as their financial counterparts—but their losses cannot be overcome by a printing press. Some losses have resulted simply from being in the wrong place at the wrong time. Student unrest in the 1960s, for example, led to a kind of "flower power" failure when technicians couldn't enter campus gene banks.[21] Irreplaceable accessions were lost. In other cases, negligence turned cold rooms into hothouses. The most faithfully reported bank failures have taken place in the Third World, where power failures are endemic.

Even collections duplicated elsewhere have problems. Two hundred of the four hundred bean samples at the University of Visçosa in Brazil were lost to an electrical fire in the mid-seventies. Much later it was learned that the duplicate collection held at Purdue University was judged by IBPGR to be of "doubtful viability."[22] In another case, four thousand bean seed accessions in Honduras and their duplicates at the CGIAR-affiliated International Center for Tropical Agriculture in Cali, Colombia, were jeopardized when germination standards were allowed to drop to dangerously low levels in both collections.

Sometimes invaluable collections disappear as soon as the original collector/curator retires from the scene. Such may be the case for Dr. Wisalk's Oslo bean bank, which does not seem to have survived the good doctor's retirement. A bean collection in Sweden seems to have vanished with the unexpected death of its keeper.[23] Important vegetable material in Holland was in urgent need of rejuvenation, according to a 1981 IBPGR study. A major corn collection was lost in transit while its duplicate was destroyed in a flood. Even the log books were washed away, leaving corn breeders to speculate on what they were missing.[24] In several other cases, corn material gathered in the 1940s or 1950s has simply been jettisoned or allowed to deteriorate by disinterested breeders.[25]

These are not isolated examples. We know more about failures in the South, and most about crops like beans and corn, solely due to the diligence of specific scientists who voiced their concern in the muted environment of IBPGR meetings. In 1979, before the subject of genetic resources was drawing much political attention, an IBPGR working group on forages became the first scientific body to address the wider problem. In its unpublished conclusions, the expert panel stepped beyond its mandate to declare, "It is estimated that even in developed countries such as [the] USA and Australia from half to two-thirds of accessions brought in over several decades have been lost."[26] To underline their concern, the experts stressed that they were "aware that enormous losses of valuable material have taken place in past collections and are continuing . . . The problem is not confined to developing countries." If the South has electrical failures, the North has bureaucratic ones.

For reasons beyond our comprehension, this bombshell went totally unnoticed. In 1983 and again in 1984, William Brown, past chair of Pioneer Hi-Bred International and a member of the IBPGR forage crop working group, warned American audiences that in at least one crop, "We are losing more genetic diversity in the banks than we are in the fields."[27] Still no one noticed, and IBPGR continued to promote gene bank storage to the exclusion of complementary strategies.

As IBPGR executive secretary, Trevor Williams actively discouraged publication of a complete study of *in situ* conservation options prepared in 1980–81 by Robert and Christine Prescott-Allen, even though the study had been contracted by IBPGR. The Prescott-Allens included a brief account of some of the problems facing the high-tech gene bank. Williams personally crossed out all negative references to gene bank safety.

While a significant number of problems with gene banks are associated with mechanical or electrical failures, most, as the IBPGR Forage Crop Working Group was careful to note, are human problems. Perhaps the clearest examples of the human side of the problem can be found in the United States, largely because that government has taken the trouble to document publicly mistakes that many other governments might opt to hide.

The first indication that U.S. gene banks were in trouble came in November, 1979, when a CGIAR review team visited the National Seed Storage Laboratory (NSSL) at Fort Collins, Colorado—possibly the world's largest and most important gene bank. Even as Swaminathan was raising the specter of genetic politics at FAO in Rome, the review team was horrified to find the international storage of seed in a total shambles. Discreetly confining their comments to terms like "warehouse" conditions, review team members nevertheless made their opinions known to U.S. government officials.[28]

Their alarm led to an in-house study undertaken by the U.S. Department of Agriculture and published in October, 1980, which warned that the "levels of funding and activity for this work are simply inadequate. . . ."[29]

Six months later, a special report by the U.S. General Accounting Office blew a long but very low whistle on the whole U.S. gene bank system. The government's conclusions are reported here in their entirety:

> The present system does not comprehensively address the real risks of genetic vulnerability. Potential crop failures are a national and international concern, and the regional efforts have not added up to an effective national program. Critical policy questions have not been addressed, indications are that germplasm protection and preservation mechanisms are inadequate, and comprehensive plans have not been made to cope with present and future problems. The system's organizational structure cannot sufficiently address these problems, and the Department's recent changes to germplasm management are unlikely to solve the problems.[30]

Predictably, the government's own investigators found that the government failed even to "adequately perform the housekeeping chores of collection, maintenance, and evaluation of germplasm stock."[31]

The low whistle went unheard or unheeded until that autumn when Ann Crittenden of the *New York Times* offered a scathing description of the Fort Collins facility. "In this innocuous and unguarded facility," she charged, "subject to power failures and so crowded that the seeds are piled on the floors in brown cardboard cartons and sacks, the germ plasm on which all global agriculture is based is supposed to be preserved forever."[32]

Even these harsh critics missed the more subtle problems. When K.L. Tao of IBPGR visited Fort Collins in the spring of 1984, he was deeply concerned that at the then-current rate of sample acquisition, the bank would be officially full by 1988. He was still more concerned, however, at the way in which seeds were being dried. IBPGR recommends that seed be dried at about fifteen degrees centigrade over a period of several days. NSSL staff were placing the seed in ovens, jacking up the temperature to thirty-six or thirty-eight degrees, and leaving the samples to cook for as long as twenty-four hours. Wrote Tao, "*the drying temperature in NSSL is too high.*"[33] (Tao underlined this comment in his travel report to Williams.)

Tao had other comments: Louis Bass, then NSSL director, claimed that the germination standards were far too high. Tao was worried that the Fort Collins standards were too low. Of particular concern was the size of samples. Tao observed, "*There is an urgent need to increase the sample size of these accessions.*"[34] (Again, his emphasis.)

Undoubtedly improvements have been made. The National Seed Storage Laboratory has a new director, Dr. Steve Eberhart, and the government's germplasm system a new boss, Dr. Henry Shands, who came to the job from the corporate world. Shands, a quiet but committed man, has pushed through a number of improvements and has earned the respect of both critics and supporters of the system. Despite this impressive turnabout, more needs to be done. To receive the necessary financial support, the system will have to become publicly self-critical. Unfortunately, Shands and others will have to talk as much about the system's shortcomings as its improvements if they expect Congress to provide more money.

William Brown of Pioneer Hi-Bred thinks the system has fewer problems now than a decade ago. Certainly the U.S. government has been unstinting in its reports. Yet another study, this from the Office of Technology Assessment of the U.S. Congress, began coming out in sections in 1986. Not to be outdone, the U.S. National Academy of Sciences launched its own fresh round of investigations in 1987. Were the rate of reporting indicative of the rate of improvement, we might be more optimistic.

Brown may have been comparing the American system with that in Canada. In 1976, the Canadian government seriously contemplated asking IBPGR for foreign aid to help support its national gene bank. The idea was nixed in fear of diplomatic embarrassment.[35]

In the years following, the Canadian public grew increasingly concerned for the safety of seeds in the bank. Bank curators made some improvements and stoutly defended the quality of their facility, declaring in mid-1986 that national and international stocks were perfectly secure. Short months later, a reporter using Canada's Freedom of Information Act discovered that even as the bank's newsletter was trumpeting the safety of its seeds, bank officials were exchanging letters warning one another that the world oat collection was slowly deteriorating. "The situation is now reaching critical proportions," one insider advised, adding that one or more of the international collections might have to be sent elsewhere for safekeeping.[36]

Dr. Eric Roos, a plant physiologist at NSSL in Fort Collins, drew the alarming conclusion for Ann Crittenden: "Within five to ten years of storage, you may have lost half the genetic material you started with."[37] Most of the messengers delivering the bad news were coming from within the system, it seemed. Yet Bill Brown and assorted government officials still reserved their harshest criticisms for outsiders who tried to draw attention to the problems. Opportunities to build bridges and mobilize popular support for improvements to the system were squandered. For repeating the conclusions of government reports, potential allies were accused of fear-mongering. At this rate, it seems we will soon have nothing left to fear but fear itself.

Seeds are placed in gene banks not so much to preserve seeds as to preserve diversity, the variation within populations. The two are not always the same. Seed collected in a Third World farmer's field is not likely to be of one uniform type. The typical bean field may contain a mixture, readily evident from the diversity of colors, if nothing else. Gene banks—the good ones—endeavor to preserve the diversity in the field as represented in the collection sample. It's not easy, even when everything works properly.

Eric Roos began experiments in the 1970s to determine how much diversity would be lost during long-term storage in good conditions. In time all seeds lose the ability to germinate. But they lose this ability at different rates. Furthermore, the genetic qualities that determine whether one seed will withstand storage better than another can be linked to other traits, such as disease resistance. For practical and visual effect Roos took an equal number of seeds from eight different bean varieties and artificially aged them to simulate storage. After he had repeated the aging process a number of times, four varieties lost the ability to germinate. Half the diversity was gone.

When germination rates fall too low (how low is too low varies from bank to bank) or when the size of the sample gets too small, the gene bank will remove some seeds from storage and grow them out to regenerate the sample. At the U.S. National Seed Storage Laboratory, germination rates are allowed to deteriorate thirty-five to forty percent before the sample is regenerated,[38] though the International Board for Plant Genetic Resources recommends that only a five to ten percent deterioration be allowed.[39] IBPGR's concern is warranted. Studies with lettuce, for example, have shown that an increase in chromosomal damage was "quite rapid" where germination rates had fallen only ten to twenty-five percent.[40] As seed begins to change in storage, it experiences increasing deterioration in potential growth rate and development, storability, uniformity, plant resistance, yield, and field emergence, and an increase in abnormal seedlings before the ability to germinate is lost altogether.

Regeneration can also affect the diversity of a sample, through differences in response to disease, insects, and weather; different levels of productivity; and human error—among other things. Next Roos tested the results of regeneration. After fifteen cycles of aging and regeneration, only two of the original varieties remained.[41] Six had become extinct—extinct as a result of "preservation" in the gene bank. Natural selection was taking place: varieties were becoming adapted to the seed bank. As Roos spread out samples of seeds from each cycle for a public television documentary in 1985, the diversity of colors in the original sample quickly disappeared. The last sample contained plenty of seeds to restock the bank, but they were all one color. Lots of seeds, not much diversity.

In the 1970s, U.S. bean accessions in "critical need of regeneration" were "allowed to deteriorate to very low germination" rates and very small sample sizes due to the nonrenewal of a contract with the University of Hawaii to do the work.[42]

Regeneration of seed samples is not as simple as it may seem. As indicated, climate, germination rates, and resistance to diseases and pests found at the grow-out site may affect the seeds in the sample unevenly, thus altering the composition and ratios of different types of seed in the sample (problems compounded when a number of samples are thrown together in a bulk for regeneration). Take for example the problem some university staffer or student will encounter when regenerating "Plant Introduction 229772," one of several thousand bean samples found at the U.S. gene bank. Dr. Roos found that the tan/brown fleck seeds took about forty-seven days to flower, whereas the purple/black splash seeds in the same sample took seventy-four days. When seeds are harvested could dramatically alter the genetic composition of this sample. An early frost could also reduce one type disproportionately. Imagine yourself trying to grow this batch of beans and return a sample to the gene bank that is representative of the sample you were given to grow out. Remember that the color variation is only one aspect of the diversity of the sample. Some differences can't be seen. And then imagine that you've been given not just P.I. 229772, but a couple of hundred samples to regenerate. No matter how conscientious you are, genetic changes will take place.

The dilemma is clear. If germination rates in the seed banks are to be allowed to only drop five to ten percent before regeneration, frequent grow-outs will be required, exposing the sample to inevitable losses. If, on the other hand, germination rates in the banks are allowed to deteriorate beyond those percentages in order to avoid the dangers of regeneration, then more and more diversity will be lost in the bank itself as the samples suffer the effects of storage. How much of the diversity originally collected and stored in gene banks over the last fifty years is still there? We can only guess. Roos's research shows that while technical conditions in gene banks can be improved, a certain amount—perhaps a very large amount—of genetic erosion will nevertheless take place.

Pick your poison. Keep the germination high with frequent grow-outs and accept genetic drift and losses in the collection. Or fight drift by allowing the germination level to decline, and accept gene losses that way. Seed keepers are faced with a scientific quandary that can only be solved by employing a little diversity in their methods of conservation. They also need to undertake more research into the effects of long-term storage. But the chance to pick a non-poison is closely tied to the political will of governments to support seed keeping. Dr. Roos's bank, which can and must play an important role in any good conservation plan, has been plagued with chronic underfunding for years. Problems there are due not to the personnel, but to indifferent politicians and past Washington administrators.

WHEAT CHECKS

The problems of the current system might best be shown through examination of one specific crop. At the same time as the Rockefeller Foundation and IBPGR were claiming a kind of Pyrrhic victory over the gathering of cereals, Chris Chapman, the IBPGR wheat officer, reported on the state of affairs of one of the board's highest priority plants.[43]

The Chapman study readily concedes that there have been collection shortcomings: "There are a number of possible reasons for this—some areas are still unexplored; the occurrence of a species within its general area of distribution may be sporadic both in space and time; plants may have been immature or already have shattered when a mission was in the area; and the distribution may, in part, be inaccurate."[44]

While claiming that most of the key countries in which landraces of wheat are endemic have been "well-collected," he adds that "when plotting missions in these and other countries it is evident that most collecting has been carried on in the vicinities of the major road networks." And there are a few important places which may have been missed altogether: "Areas of eastern Europe, southern USSR, and north Africa appear largely uncollected."[45] Furthermore, some areas cannot be collected even today due to "political situations."

From the deficiencies of collecting practices, Chapman turns his attention to conditions in the seed banks. For example: "One important collection, however, has major problems with regard to the accuracy and up-to-dateness of its inventory, and this needs to be dealt with as a matter of urgency."[46]

Despite all the problems of seed drying cited by Tao and the concerns for grow-outs identified by Roos, as well as the usual risk of power or similar failures, Chapman writes, "The total loss of an accession through seed aging is highly unlikely." But then he goes on to negate his own optimism.

> There are, however, two important exceptions: the Vavilov Institute at Leningrad and the National Bureau of Plant Genetic Resources (NBPGR) in New Delhi. Both store their seed in ambient room conditions, and so need to regrow their samples regularly, with all the work and hazards to the germplasm that implies. In the case of the Vavilov Institute full duplication of the collection in long term store at Krasnador would be a solution. To date this has only been done for a third of the collection. For the NBPGR collection a new store is to be

> completed. Nevertheless duplication is still desirable in view of its comparatively small size (c. 1100 accessions).

Having attacked the safety of storage of the world's largest wheat collection (at the Vavilov Institute) as well as the Indian bank, Chapman later adds a devastating criticism of the most important wheat collection in Asia: "At the collection at the University of Kyoto, only small seed samples are kept and only five plants per accession are grown for regeneration. Since this is repeated at intervals of about five years the loss of within-population variation must be considerable."

With Leningrad and Kyoto in doubt, only Beri, Italy, and the Fort Collins bank—of those with world class collections—remain. But Tao of IBPGR has already criticized Fort Collins for the same sample size problems condemned by Chapman at Kyoto. And as for Beri, Chapman notes, "Against this background it is difficult to say which collections have good or bad documentation, although it must be mentioned here that the inventories of the collections at Kyoto and Beri end at 1979 and 1978 respectively."

Now, with the safety of all the major wheat collections in doubt (as well as the Indian collection), Chapman goes on to list problems with many of the key banks in the center of wheat diversity. According to Chapman there are a number of important collections still in need of duplication. He gives Leningrad and New Delhi first and second priority, and he ranks the collection in Turkey third, saying, "This is particularly important as the latest Index Seminum reports the loss of material without stating a cause." Chapman's fourth priority is material scattered in short-term stores in Iran in need of both multiplication and duplication. Finally, he expresses concern for the status of Chinese collections, noting that "the extent of this need is simply not known." In other words, we cannot have confidence in any of the major wheat collections.

After reviewing all of the shortcomings in the collection, storage, and documentation of wheat, it would hardly seem imperative to attempt an estimate of the number of wheat accessions "safeguarded" in seed banks. Nevertheless, Chapman enthusiastically reports, "The amount of cultivated wheat germplasm in collections is very great indeed, the number of unique landrace accessions exceeding 60,000. By comparison little collecting remains to be done."[47] (One month earlier, the Rockefeller Foundation scientists had declared that one hundred thousand distinct wheat landraces were in storage with ten percent still to be gathered.[48])

Chapman concedes that the figures are confusing. Later in the study he explains, "Despite three years of work by the wheat officer, and largely because of the incompleteness of the data available to him, the amount of wheat germplasm in collections is still unclear . . ." With this final pathetic observation, Chapman recommends that wheat be downgraded on the list of crops needing collection to "priority 2." And for one of the world's best collected and protected crops, Chapman observes, "this would seem to be an adequate resource for breeders *for the foreseeable future*" (our emphasis).[49]

SOME ADDITIONAL OBSERVATIONS ON THE SYSTEM

Many—but not all—of the problems in current conservation strategy boil down to a shortage of money. This in turn indicates a lack of political will. The cost of the high-tech system involving international collection expeditions, overseas shipment of germplasm to distant gene banks, and the costs of maintenance and rejuvenation are enormous. When IBPGR began collecting in 1974, it was possible to gather a sample of African rice for about one U.S. dollar. By 1980, some crop accessions were coming in at $400 per sample.[50] By the mid-eighties, a conservative estimate of the average cost (there are no hard figures from IBPGR) would be $42 for an easy-to-collect cereal sample.[51]

Once collected and banked, the seeds have further expenses. The cost of growing out a single sample can run as high as $50 to $200 per accession.[52] Even the ongoing passive storage of seed a decade ago was estimated to run between $1.89 and $10.74 per sample per year.[53] If present material were being correctly conserved, the annual costs for two million accessions would be at least $50 million per year.

Figures like these have led Trevor Williams and his board to speculate on yet more complex technological alternatives. Speaking before a joint FAO/IBPGR/UNEP technical conference on plant genetic resources held in Rome in June, 1981, Williams waxed almost poetic about the energy-saving potential in what scientists call "naturally-occurring cold environments" (read "Antarctica") and the energy efficiency and safety of long-term storage in the seed equivalent of Skylab.

The mind boggles. A 1982 report to IBPGR on some of these ideas admitted to a few problems. The ad hoc working group conceded that

Antarctica would indeed keep seeds cold, but there would be the problem of keeping the technicians warm. Then there was the energy and cost involved in going to the ice box to get the world's breakfast. Finally, the group identified a rather serious shortcoming: "ice movement."[54]

Third World delegates to the conference viewed Williams' dreams with understandable skepticism. Argentina has been trying to get IBPGR support for a no-ice low-energy bank in its own arid south without any luck. Space disasters raise the possibility that the launching of T.T. Chang's life's work into orbit might be a short ride.

Not that safety is easy to find. After building their main gene bank near Lund in southern Sweden, the Nordic countries decided it would be prudent to store a back-up sample of everything in the permafrost of the high Arctic. They had the ideal place, Spitsbergen, just eight hundred miles from the North Pole and under the trusteeship of Norway, but "international" territory to the forty countries that signed the Treaty of Versailles. The seeds have been stored in wooden boxes nine hundred feet down the shaft of a working coal mine.[55]

Cheap—but not as safe as the Scandinavians had hoped. As part of the Svalbard archipelago, Spitsbergen lies astride the Arctic artery that links the world's largest naval base at Murmansk with the Atlantic Ocean. As Kevin Done of the *Financial Times* has commented, "two-thirds of the Soviet submarine-based strategic missiles and about 50% of Soviet attack submarines" come to nest at Murmansk.[56] Twenty-five miles east of the gene bank is a Russian mining settlement—also part of the Norwegian trust but, due to the treaty, fully able to operate on Spitsbergen. The region has become a major strategic location for military surveillance.

In the final analysis, the conservation of plant genetic resources is and always has been a political issue. In this, Third World countries have far from cornered the market on purity or integrity. But the seeds come from the South and their storage there would be less costly and closer to home for the people who need them most—poor farmers. In the next two chapters we will explore the mingling of the politics and the science. And we will see why the safeguarding of our daily bread increasingly requires not just good science, but good politics as well.

CHAPTER NINE

Politics of Genetic Resource Control

[Our] national interests are dependent upon continued access to the world's germplasm.

—American Seed Trade Association

The politics of plants are not new. Humanity has long squabbled over the control of the fruits of the earth. As Third World governments grew to understand the danger to genetic resources and the economic advantage to those who control them, a struggle began in the United Nations Food and Agriculture Organization (FAO) to assert authority over the International Board for Plant Genetic Resources (IBPGR), and to lay out ground rules that would ensure "exchange" of germplasm among all countries.

On a drizzly Monday evening in November of 1981, raincoated ambassadors in Rome were hastening out of FAO's Building A into the dismal night to contact their home countries and consult with their colleagues. A diplomatic furor had erupted in the Red Room over a Mexican resolution on the control and exchange of plant genetic resources. The resolution had caught all save the Latin Americans unprepared. Naked without their briefs, ambassadors still sensed that the issue was important—especially given the stony response of industrialized countries in the conference.

But as the day ended in Rome and telex machines disgorged the Red Room's business into in-baskets back home, James L. Buckley, U.S. Under Secretary of State for Security Assistance, had the situation well in hand. Standing before a Washington, D.C. crowd of three hundred, Buckley opened the U.S. Strategy Conference on Biological Diversity. He took the high road, espousing his concern for the snail darter and warning his audience that they were dealing with "an international problem requiring an international approach."[1]

It was for Dr. Jim Murray of Chicago's Policy Research Corporation to deliver the heavy political goods. In a tough-minded speech Murray pointed out, "The importance of biological diversity to the future of genetic engineering cannot be overemphasized. Germplasm is the fundamental resource . . ." and, Murray stressed, "developing countries will have an advantage in that they are the sources of a large percentage of the germplasm resources of the world."[2] Access to germplasm he described as "the limiting factor" facing genetic engineering companies. Murray called for a "context" in which industrialized countries and developing nations would negotiate for the sharing of benefits, and he repeatedly warned that the Third World would become aware of the value of its raw materials.

Jim Murray also cautioned the genetics supply industry. "Some companies treat their germplasm as a proprietary resource," Murray advised; ". . . secrecy is already being extended to private germplasm collections . . . [Corporate interest] will present obstacles to widespread collection, management, or sharing of germplasm." While Murray saw opportunities to persuade corporations to help with genetic resource conservation, he warned that unless strong public policies were adopted and actively pursued, the involvement of "the private sector is more likely to be a hindrance than a help."[3]

At the beginning of the 1970s, the Americans had learned about the cost of uniformity with the collapse of their corn crop. The Russians got the same message at almost the same time with the collapse of wheat. With close to fifty biotechnology companies banging at the doors of the world's gene banks by 1981, Murray's words underscored the profitability of germplasm.

Among company representatives at the State Department's conference and among their counterparts in Europe, Australia and Japan, the message was received and understood. But the politics were tricky. Thanks to two outstanding young Mexican ambassadors, both sons of former presidents, the Third World was catching on faster than even Murray had expected. Industrialized governments found themselves talking about germplasm as the "common heritage" of all humanity at FAO battles in Rome while enacting seed patent legislation and counselling corporations to monopolize this common heritage at home. Indelicate diplomacy, at best. To see how matters reached this pass, we need to revisit the tombs of the Pharaohs and turn back the leaves of the green revolution.

THE BOTANICAL CHESS GAME

On a scalding June day in 1976, we rented bicycles at Luxor and crossed by ferry to the west side of the Nile at Deir el-Bahri. Pedalling the well-travelled macadam road through Egypt's desert sun, we searched for one special temple among the tombs and ruins that crowd the rocky landscape. Queen Hatshepsut's temple was not hard to find. Standing elegantly at the foot of towering sandstone cliffs, the temple faces east toward the Nile. Built around 1480 B.C., it features colonnades depicting the great queen's military expedition to the Land of Punt (East Africa). As the reliefs display even today, Queen Hatshepsut had launched one of the world's first plant collecting expeditions.[4]

Seed or plant gathering has been part of civilization's military repertoire for most of recorded history. A thousand years before Hatshepsut another pharaoh, Sankhkere, sent an expedition to the Gulf of Aden seeking cinnamon and cassia. Emperor Sheng Nong, meanwhile, was gathering the world's first medicinal plant collection around his palace in China.[5] A few centuries later, Nebuchadnezzar built his wife the Hanging Gardens of Babylon, stocked with exotic flora from around the Middle East.[6]

But it took Hatshepsut's expedition to make plant collecting a fully integrated function of every military campaign. Records show the capture of plants from as far away as Syria. A shipwreck off the coast of Turkey has been dated back to this period. The vessel was laden with precious booty—gold, bronze, silver—and seeds. Even in the days of the pharaohs, some seed varieties had a commercial value far beyond their weight in gold.[7]

The attempt to monopolize seeds or food is as old as history. In the seventh century B.C., the Syberites granted cooks monopoly over their recipes for one year.[8] Roman emperors, on the other hand, opposed the monopolization of fish.[9] Today, the combined effects of genetic erosion and the development of genetic engineering make the monopolization of germplasm a global political concern.

During the Middle Ages in Europe, wondrous tales (mostly false) were told of plots to steal tulip bulbs from the Court of Suleiman the Magnificent and to smuggle coffee beans in hollowed-out walking staves carried by monks returning from Palestine. In the 1600s, Europeans awoke to the realization that, far from being scattered evenly about the globe, the fruits of nature are concentrated in the tropics and subtropics—in exactly the places where Europe is not. And English explorers set out to steal from the botanically rich and give to the genetically poor—themselves.

The pace of British plant introductions can always be linked to the spread of political power. Historian Kenneth Lemmon directly relates the increase in plant accessions to specific events such as Britain's takeover of the Cape of Good Hope from the Netherlands, the colonization of New South Wales and Nova Scotia, and, most of all, the Treaty of Utrecht provisions that gave England the right to trade in Spanish territories.[10] Between 1731 and 1763, the number of exotic plants entering Britain doubled. On the eve of the coronation of Queen Victoria, people in the British Isles were growing thirteen thousand species of exotic flora.

The decorating of the English country garden grew to a major undertaking. Whereas fewer than a hundred plants were introduced to the British Isles in the sixteenth century, close to a thousand arrived in the seventeenth and almost nine thousand exotic plants were brought in during the eighteenth century.[11]

The domestic impact was spectacular. Diet improved directly with a host of new food plants such as potatoes and tomatoes, and indirectly, with the introduction of new pasture plants from Africa and Latin America.[12] The new plants had an impact in the textile industry and in every facet of the fledgling chemicals business ranging from paints, dyes and resins to medicinal preparations. A substantial commercial trade also developed around the provision of exotic flora for the gardens and greenhouses of Europe's aristocracy.

To an extent which has not been properly recognized, the history of colonialism is a history of the struggle to capture and monopolize botanical treasures. The European powers moved troops and fleets about the earth in order to control production of or trade in commercially important plants. With the discovery of safer means for transporting live plants, the struggle for monopoly took on the form of a botanical chess game. The major powers uprooted crops from one continent and colony to install in another for strategic advantage.

As part of Europe's campaign to redress the injustices of nature, the Dutch cut down three-quarters of the clove and nutmeg stands on the Moluccas in order to confine production to three defendable islands. In their traditional egalitarian way, the French added a botanical twist to the call for "Equality . . . or death," committing to the guillotine anyone caught stealing live indigo plants from their Antigua island stronghold—plants which the French themselves had stolen. By the time John Donne wrote "No man is an Island," every plant had an island. The transfer of spices from Southeast Asia to Africa and the Caribbean became a comic chase as feuding European interests stole, smuggled, and bargained valuable species from the Moluccas to Penang to Reunion, Ascension, Zanzibar, Grenada, and Antigua. The island-hopping also moved from west to east as the Dutch used their island base at Curaçao to break Spanish domination of the cocoa trade. Moving production to Sâo Tome, the Dutch were outdone by enterprising slaves who eventually landed seed on the African coast.

Europe's battle for botany was often backed by national embargo legislation. As early as 1556, Spain's Council of the Indies convened in Madrid to pass laws making it illegal for other foreigners to explore for plants in New World Spanish possessions.[13] Attention to such legal niceties was minimal, however, and all the European powers felt it their inalienable right to violate the national laws of other states in order to gain control of economically useful plants.

Spices and ornamentals were not the only chess pieces. In her excellent book, *Science and Colonial Expansion,* Lucile H. Brockway has done much to document the importance of tropical plants to the colonial powers. At the beginning of the seventeenth century, the Dutch transferred coffee production from the Indian Ocean crescent to Suriname and much of Latin America. The movement of cocoa from Central America to West Africa also had a profound impact. In the nineteenth century, tea was transferred from China to South Asia and East Africa, sisal was taken from Central America to East Africa, and oil palm from West Africa led to new plantations in Southeast Asia. Rubber from the Amazon joined cinchona from the Andes on new estates in South and Southeast Asia. Asian bananas became the backbone of Caribbean production. Cotton, sugarcane, citrus, and a dozen other tropical crops that had spread in ancient times from their gene centers were shifted to still newer areas of commercial production.

The effect of these botanical transfers is quite literally beyond the possibility of economic calculation.[14] The Andean republics that once controlled the production of cinchona for malarial drugs lost almost the entire market when new plantations were opened in Asia. Brazil—once the sole producer of rubber—fell to less than five percent of the global market when new production sprouted in Africa and Asia. China's loss of much of the tea trade; Mexico's loss of sisal production (for baler twine and ship rigging); West Africa's loss of oil palm (which now occupies fifteen percent of the massive multi-billion dollar vegetable oils trade)—all had staggering economic consequences for the development of the countries concerned.[15]

Some have argued that the swapping of plants back and forth across the globe has been a plus for all involved. Shipped far away from their centers of origin, many major crops did better in their new homes, safe from their usual pests and weedy rivals. If dominance in the production of one crop were lost, another newly introduced crop was sure to take its place. What Latin America gave away in rubber and cocoa it got back in coffee and citrus. It all evened out.[16]

Convenient as it may be, this kind of reasoning belies the real impact on Third World people at the time, as well as later. Cinchona workers in the Andes and rubber workers in the Amazon did not merely come back after a weekend lay-off and switch merrily to coffee and citrus groves. They suffered considerable dislocation.

Further, some countries and continents were clear winners or losers. Asia was a winner. Africa was a loser. But the ultimate winners were the colonial powers in Europe and North America. The surplus value of all the new and increased production accrued to those who orchestrated the plant transfers. Wherever the plants ended up, the money was safely routed to London, Paris, and other capitals of the North.

A longish trolley ride and walk from the Amsterdam railway station takes you to the stone entranceway of the old botanical garden. When we last visited it in 1982, the garden was celebrating its three hundredth birthday. Cramped and crowded, it was clearly showing its advanced age. But, tucked away in a battered greenhouse without any sign of distinction, gardeners showed us the living remains of an ancient coffee tree. Not just any coffee tree—*the* coffee tree.

Coffee, as already mentioned, is native to Ethiopia. Arab or Persian traders moved the bean to Yemen for cultivation at least a thousand years ago and only seven seeds were later sent to India to launch production there and in Sri Lanka (then Ceylon). When the Dutch took over Sri Lanka from the Portuguese, they also inherited coffee. One tree was replanted in Indonesia (Java) and a cutting from that tree was shipped safely to the Amsterdam Botanical Garden in 1706. The first healthy progeny from this tree were sent as a gift to King Louis XIV of France, who tended them in a specially constructed glasshouse at the celebrated Chateau Marly in Paris.[17]

Nine years after its arrival in Holland, cuttings from the Amsterdam tree were packed off to the Dutch colony of Suriname. Other cuttings from King Louis' Jardin Royale were shipped across the Atlantic to Martinique in 1723.[18] And thus it is that the entire coffee industry of Latin America is based upon seven plants taken from Yemen a thousand years ago, and thence one plant in Java less than three hundred years ago.

The central biological and economic factor missing from colonial histories is embodied in the travels and tribulations of coffee. One tree represents the genetic base for the entire Latin American coffee industry. And coffee is not alone. Four Nigerian palms shipped to Indonesia around 1848 comprise the entire genetic make-up of the Asian oil palm industry. Twenty-two rubber plants gathered from the same stand in the Brazilian Amazon in 1876 are the basis for almost all Asian rubber production. The sisal trade in East Africa hangs upon sixty plants stolen from the Yucatan Peninsula at the dawn of the twentieth century. Among the plantation crops that were uprooted and given new centers of production in the colonial era, only tea (taken from the Chinese following the Opium Wars around 1848)[19] and cinchona (illegally obtained from the Andes through several expeditions during the period 1854–65)[20] were transferred with a somewhat wider breeding base.

Such extreme genetic uniformity usually translates into genetic vulnerability. The risk of crop loss threatens the livelihood of tens of millions of small producers in Asia, Africa and South America. To defend themselves and their crops against the disastrous consequences of genetic uniformity, farmers are driven to use costly chemicals that endanger health and profitability, and damage the environment. Whole national economies are exposed to potential disaster from a virile new coffee rust in Central America, black sigatoka in banana, or the sudden appearance of a mutant canker in citrus groves. The legacy of the colonial transfers continues to haunt the South today.

The chess game could not have been played had it not been for two great developments, one scientific and the other institutional. In 1829, the invention of the Wardian case allowed the relatively safe transfer of living plants from one corner of the globe to another.[21] With the aid of the case (actually a terrarium), British botanists were able to transfer six times as many plants in fifteen years as they had in the preceding century. Without the case, rubber, cinchona, sisal and tea would never have survived their great journeys.

The institutional development was the spread of a thin green line of botanical gardens girding the globe in the tropics from Havana to Trinidad, Manila and Vietnam, on to Bogor (Indonesia), Singapore and Calcutta, Colombo (Sri Lanka), and onward through Mauritius and Entebbe (Uganda) and across to Rio de Janeiro.[22] From these strategically located institutions, a small band of altruistic botanists went forth and collected, analyzed, and forwarded—to counterpart institutions in the North—the world's flora, both beautiful and beneficial.

The great knowledge they gained is available to any scientist today, but the profits these plants made possible have long since accrued to the early botanists' Northern homelands. Seldom did plants move simply from Africa to Asia without going first through "parent" gardens in London, Berlin, Amsterdam, or Paris.[23] Almost as a by-product of their scientific pursuits, botanical gardens utterly transformed the agricultural economy of the world.

Not that Europeans "invented" botanical gardens. From the Hanging Gardens of Babylon onward, the nobility of great civilizations have gathered about them the beauty of ornamental plants and trees. When Cortez overthrew the Aztecs in 1519 he discovered a vast botanical garden—acres of exotic plants sown together from throughout the Aztec Empire and beyond—attached to an agricultural school with advanced methods reputed to have surpassed those of Europe.[24] By contrast, the Amsterdam garden potted its first plants in 1682 and London's Kew Gardens didn't blossom until almost a century later.[25]

But the European colonizers went a step further, each organizing a strategic network of botanical gardens in the service of science and industry.

BACK TO THE FUTURE

Today's equivalent of the Wardian case may be the gene bank and cell library. The technological concern is no longer safe transfer of living plants, but transfer and safekeeping of germplasm or microorganisms essential to the current agricultural revolution offered by genetic engineering. The new thin green line is composed of the international agricultural research centers (IARCs) of the Consultative Group on International Agricultural Research (CGIAR) system and the smaller and more informal MIRCENs (microbiological research centers) of United Nations Educational Scientific and Cultural Organization (UNESCO) scattered throughout the globe.[26]

Like the tropical botanical gardens that preceded them, the IARCs and MIRCENs (of which there are over twenty) of today have their "parent" centers. The species-specific centers are the operational core, but they are backed by a growing string of policy centers located in such places as Washington, The Hague, and Rome. In the 1980s, few of the Northern centers had the political muscle of an IRRI or CIAT, but their role is growing and their influence is being felt. For the green revolution IARCs, the real metropolitan control rests with its donor state members, CGIAR, and the World Bank where CGIAR is headquartered.

The gene revolution MIRCEN structure is more informal and operates with the benevolent support of UNESCO in Paris without the kind of financing (or controls) applied to IARCs out of Washington. But the effect remains the same. The yeasts, fungi and bacteria of Third World soils, swamps and savannas are found, catalogued and exchanged. Again, the information is available to all—but the North is in a position to use the new knowledge immediately and most easily.[27]

The tide of exploration that once crested about the old botanical gardens has now ebbed. In Britain, the current now runs to Cambridge's National Institute of Agricultural Botany instead of Kew Gardens and the French prefer to send their genes to their tropical research institute at Montpellier rather than to Paris. Instead of the Berlin gardens, the Germans now have the Max Planck Institute and an old Luftwaffe base-turned-gene bank near the East German border. The Dutch have moved from Amsterdam to new digs at the agricultural university at Wageningen.

The full role of the MIRCEN network remains uncertain. The importance of the IARC network in the supply of plant genetic resources, however, is very clear. Much has been said about the green revolution.

Many observers consider its initial phases to have been counterproductive. But its advocates today claim that the scientific community has learned from its mistakes and is now contributing to Third World development with much more "sympathetic" tools. True or not, it is a fact applauded by some and dreaded by others that the IARCs are continuing to draw peasant farmers into the mold required by the western food system of the developed world.

Studies of the IARCs have tended to focus on their impact on the South. Very little has been said of their impact on the North. Like the botanical gardens before them, the IARCs of today are making a substantial contribution to the transfer of biological treasures to industrialized countries.

The IARC contribution comes on two broad fronts. First, the international centers identify and transfer large quantities of "raw" and "improved" germplasm which may ultimately appear in farmers' fields in industrialized countries. Secondly, the centers have become a major training ground not only for scientists from the South, but also as a "finishing" school and research laboratory for scientists from the North. Although accurate calculations of the value of these two services are immensely difficult to make, a giant and hitherto ignored benefit has indisputably accrued to Northern countries.

Of greatest importance is the transfer of germplasm. CIMMYT in Mexico supplies 127 countries with improved breeding material through its large nursery trial program. Participating stations are obliged to report back on "grow-out" results but they are welcome to keep the germplasm. Over a quarter of all the wheat nursery trials take place in the North. There are seventeen trials for wheat in countries as far north as Finland, Norway, and Sweden, and another thirty-seven in Canada.

The economic payoff to the North of access to CIMMYT wheat germplasm is spectacular. The farmgate value of the U.S. wheat crop in 1984, for example, was in excess of eight billion dollars. The American government's own estimate is that CIMMYT material was responsible for almost two billion dollars of that value.[28] That same year, the U.S. Agency for International Development (AID) gave its regular grant of six million dollars for CIMMYT work on cereals.

What is the actual value of exotic wheat germplasm to the United States? No one knows for sure and some of the data are even contradictory. Everyone agrees, however, that the value is extremely high. In 1982, the Organization for Economic Cooperation and Development (OECD) placed the figure at five hundred million dollars per annum. A year later,

American and Canadian breeders became alarmed that a disease problem in CIMMYT's test plots might lead to a temporary curtailment of germplasm transfers. CIMMYT material was used in about seven percent of the U.S. wheat crop in 1969; by 1984 the contribution had risen to over fifty-eight percent.[29]

Everyone can also agree that the North's "payment" for this economic benefit is modest indeed. The six-million-dollar AID annual grant to CIMMYT also covers corn. Given the importance of this crop to such nations as the U.S. and France, it is surprising that very few of CIMMYT's nursery trials are in the North. Yet, the North's interest is actually considerable. CIMMYT has a problem growing out corn at its Mexican gene bank, lacking personnel, time, and land on which to rejuvenate accessions. One American company has stepped into the breach and has volunteered to multiply this exotic material at its own Florida research station. That company is Pioneer Hi-Bred—the world's largest corn breeding corporation and the dominant company in both Europe and North America.

In general, North America's dependence upon tropical exotic corn germplasm has been low due to its proximity to the Mesoamerican gene pool. One North Carolina researcher, Dr. Major Goodman, estimates that only four percent of U.S. corn acreage contains any non-U.S. germplasm, meaning that much less than one percent of the total value of the crop can be traced to the Third World.[30] Nevertheless, Goodman suggests that private companies believe their demand for Third World corn genes will grow, rising to as much as thirty percent of market value over the next several decades.[31] The dollars involved are not modest. Even in 1984, tropical corn germplasm contributed at least twenty million dollars to the American crop. Were the thirty percent figure reached, the farm gate share would run to more than six hundred million dollars.

Along with wheat and corn, the third pillar of the green revolution, rice, has also led to benefits for Northern breeders. It is grown extensively in Europe's Po Valley and in the southern U.S. IRRI-derived semidwarf material was sown over 182,000 hectares in the U.S. in 1984—amounting to almost sixteen percent of the entire U.S. crop. AID predictions are that the area based on IRRI germplasm will increase substantially in coming years.[32]

Again, the dollar value of the Third World's donation (through IRRI) cannot be ignored. AID's annual grant to IRRI is the same as for CIMMYT—about six million dollars.[33] Yet IRRI contributes to $176 million of the farm gate value of the American rice crop, and even this may be a serious underestimate, since it does not include germplasm forwarded to the U.S. from T.T. Chang's 80,000-accession rice gene bank also at IRRI, a collection partially duplicated at Fort Collins.

There is no reason why the North should not benefit from research it has financed in the South. For CIMMYT to refuse to make advanced germplasm available to the North would be to violate the widely shared principle of full and free exchange.

We are concerned about another possibility, however. Because 282 of CIMMYT's wheat nursery trials are conducted in countries that allow plant patenting, CIMMYT material might be directly or indirectly used in "Northern" varieties or hybrids under private corporate control. These countries are under no obligation to share their discoveries with CIMMYT as freely as CIMMYT has shared its research with them. Our concern increases in the context of genetic engineering where the access to germplasm can be extremely important. Again, it is unlikely that the economic benefits derived from use of this germplasm in the new technologies will flow easily back to the South.

The kind of phenomenon that galls many Third World agronomists and diplomats (and even the IARCs themselves) is this: a wheat variety called CB-801—a thinly disguised version of IR-8—has been patented and is now being marketed in the United States by a private breeder (Farms of Texas Co.).[34] This is not an isolated event. Agricultural Genetics Company has lifted an important gene from a West African cowpea variety developed by International Institute for Tropical Agriculture (IITA) in Nigeria and patented it in the U.K.[35] No royalties accrue to IITA or to the African farmers who nurtured the germplasm in the first place.

As indicated, the IARCs also play a major role as training centers. In the dozen years from 1966 to 1978, for example, CIMMYT trained 886 future agronomists, virtually all of them from the Third World. But another 840 scientists—many either from the North or part of the brain drain to the North—visited CIMMYT during the same period to use its laboratories and fields for less formal programs that ranged from a week to several months. No breakdown of the nationality of all these researchers is readily available, but in our own discussions with agronomists from Finland to Australia, it is quite common to encounter women and men with practical CIMMYT experience.

In the case of IBPGR, financial support has long been given to the University of Birmingham (U.K.), for a genetic resources training program. The board has both funded Third World trainees and subsidized the overall cost of the course. Fully thirty-nine percent of all trainees have come from industrialized countries. Thus, trainees from the North have also been directly subsidized by IBPGR.

Plant breeding and genetic engineering need genes, and they need scientists to work up the exotic material into catalogued characteristics. The international agricultural research centers, often unwittingly or unwillingly, find themselves becoming the servants of private interests.

A study for Canada's International Development Research Center (IDRC) noted in mid-1983 that there have been "several instances" where IARC-developed material was acquired and patented by a private concern.[36] In informal discussions, IARC scientists tell stories of visits from corporate breeders who—short months after a stroll through IARC test plots—would announce the introduction of a new variety wholly or heavily dependent upon the IARC's material.

The late Dr. Glenn Anderson of CIMMYT was particularly outspoken in his criticism of the use private companies were making of CIMMYT. He told us one almost unbelievable story of having displayed new varieties at an exhibition in Chicago. A short while later, an American company (Anderson never revealed its name) wrote to him asking for a sample of the variety. Anderson complied. Not long after, the company wrote back asking the CIMMYT scientist for help in completing some attached forms. The forms were an application for a plant patent in the United States for CIMMYT's variety.

The first phase of the green revolution taught green revolutionaries that it is better to release improved germplasm than finished varieties. Increasingly, they see their role as supporting national breeding programs. To that end, they identify and circulate material which may have value to any interested party. In making this shift, the IARCs have responded to the desires of many Third World countries and the hopes of others that the release of advanced breeding lines (as opposed to finished varieties) might contribute to increased varietal diversity. This is as it should be. But the danger—now being recognized by the IARCs—is that they will be relegated to the role of doing basic research for the benefit of private companies. The companies can take IARC material and exploit it for their own commercial purposes on a global scale. Many IARC scientists find this possibility deeply disturbing.

On a flight to the U.S. recently, a senior IRRI official was confronted by an executive of a major crop chemical concern developing hybrid rice for

Asia. Straight out, the executive wanted to know what it would cost to buy IRRI.[37] While the IRRI official told the story for its humor and irritation value, the point was still there. IRRI is not for sale; but it may be for hire.

In the warm afterglow of Norman Borlaug's Nobel Peace Prize, the IARCs and their green revolution were relatively recession-proof. Now that the glow has faded, U.S. and European legislators are likely to see IARC research into rice and wheat as unwarranted challenges to their exports. By the mid-1980s, CGIAR's budget was on hold and real declines were universally predicted. Already the perplexed provider of germplasm, the CGIAR system, could easily fall victim to the "ten percent phenomenon." Corporations top off research budgets with an additional ten percent, but in the process gain subtle control over the entire program. The trap is almost unavoidable.

As valuable as the gift of a new IARC variety may be for exploitation in the North, genetic engineering is shifting interest from whole plants to specific genes. We have already referred to Pioneer Hi-Bred's role in testing CIMMYT's exotic corn material. This is not a unique situation. Indeed, the U.S. government now regularly turns over its exotic germplasm for grow-out help from American companies. U.S. officials describe the support provided by the companies as a "generous offer" and have even said (with some confusion), "If the U.S. Treasury had notified citizens of a gold shortage, how many would have answered the call with a direct donation?"[38] In fact, the reverse applies. The National Seed Storage Laboratory at Fort Collins—a genetic Fort Knox—is offering to give its gold to private companies for sampling. The companies have answered the call.

INTERNATIONAL LAW OF THE SEED

The stir begun by the Mexican delegation on the banks of the Tiber in November, 1981, was bound to spread. By the end of November, despite strenuous opposition from Washington and London, an FAO resolution had been passed calling for an international convention on plant genetic resources as well as a new "international gene bank." Two years later, at the next biennial FAO conference and following a series of smaller skirmishes, Ambassador José Ramón Lopez-Portillo of Mexico was able to force a favorable vote on the formation of the International Undertaking on Plant Genetic Resources as well as the creation of an

International Commission on Plant Genetic Resources. Two years after that, in November, 1985, the commission had already held its first meeting and proposals were under way to rein in the ubiquitous IBPGR and establish a world gene fund. By the time of the commission's second meeting in March, 1987, the body had a working secretariat and a clear plan of action to assert intergovernmental discipline over genetic resources. As the insightful Pakistani delegate, Javed Musharraf, put it, "Every train needs an engineer."

Understandably, the kind of diplomatic guerilla warfare waged in the lobbies and salons of an FAO conference can seem absurd and irrelevant when seen from the plateaus of Ethiopia or in the gene-splicing laboratories of San Francisco. Nevertheless, the International Undertaking on Plant Genetic Resources does—for the first time in history—lay out ground rules for the sharing of genes. More than that, it puts industrialized countries on notice that private monopolization of genes and use of genes as political bargaining chips lie beyond the pale of international acceptability. The new body offers a yardstick by which national governments and the world community can measure national and global initiatives.

The Commission on Plant Genetic Resources is no less important. Again, for the first time ever, an intergovernmental body has been created capable of applying not only scientific but also social and political realities to the collection and exchange of genetic resources. The world's donors of germplasm finally have a voice. Within five years of the passage of resolutions creating the FAO Undertaking and Commission, 114 countries had formally joined one or (usually) both structures. Notable for their absence were the United States and Canada.

The political momentum brought to bear over the battles in Rome will ultimately lead to a multimillion-dollar fund for national and international germplasm conservation. This should be new money but not "foreign aid" money. If some governments have their way, the seed industry will be taxed by national governments on behalf of the world community to pay for the preservation of genetic raw materials. Also on the horizon, at the initiative of countries including Spain, Ethiopia and Costa Rica, the UN is on the verge of creating a truly international network of gene banks, the contents of which will be legally identified as the property not of governments, but of the United Nations.

The drive toward these goals has been much more than a discreet diplomatic dance. No one who sat in the FAO plenary session as we did in

November, 1983, will ever forget the emotion and the anger as Third World delegates repeatedly overruled the FAO chairman, U.S. Agriculture Secretary John Block, when he attempted to scuttle a resolution creating the new commission. Block asked for a show of hands on the resolution. Without stopping long enough for a count, he ruled that the Third World position had been defeated. It was not a quiet meeting. Ambassadors were shouting, diplomats were darting about the room, hands were thrown up in frustration. Another show of hands—and again Block ruled that the Third World had lost. Finally a vote count was forced. It wasn't even close. The victory went to the South. Lopez-Portillo was all but hoisted onto the shoulders of his fellow ambassadors.

Although much of the debate in FAO stuck to the high road of "common heritage" and the need to safeguard the destiny of the food system, gnawing at the guts of ambassadors on both sides of the battle was the role and future of IBPGR. The more Third World delegates looked into the structure, finances, and decisions of the board, the more interested they became.

The real "donors" to the IBPGR system, ambassadors learned, were the Third World countries themselves. The South gave IBPGR fifty-nine percent of its plant collectors and sixty-one percent of the agricultural institutions supporting IBPGR's work came from developing countries. More significantly, ninety-one percent of all the samples and duplicates collected and distributed came from Asia, Africa and Latin America. Despite this, only fifteen percent of these samples have so far gone to developing nations. Eighty-five percent were distributed more or less equally among the northern-influenced IARCs and industrialized countries themselves.

The United States swallowed the lion's share with more than a quarter of all the samples. Further, the so-called "global network" touted by IBPGR appeared to be a northern network. In its annual report for 1987, sixty-seven of the designated base crop collections were assigned to western industrialized countries. Another four were identified for Eastern Europe, twenty-six were listed for CGIAR institutes, and only twenty-two were assigned to the entire Third World—one fewer than the number earmarked for the United States.[39]

Even this underestimates the economic value of the IBPGR connection. The U.S. manages 67 percent of the world's stored peanut germplasm, 24 percent of sugarcane and 20 percent of pea material, making it the lead nation for all three crops. The United States ranks second in the world for its share of chickpea and bean germplasm (21 percent and 15 percent

Table 4
U.S. Participation in IBPGR's Global Network of Base Collections

Crop	Categories	Scope of collection
Amaranth		Global
Beans	*Phaseolus* Cultivated	Global
Cowpeas	*Vigna*	Global
Eggplant		Global
Grasses	*Cynodon*	Global
Grasses	*Paspalum*	Global
Grasses	*Pennisetum*	Global
Legumes	*Zornia*	Global
Legumes	*Leucaena*	Global
Millets	*Pennisetum*	Global
Okra		Global
Onion	*Allium*	Global
Sorghum		Global
Soybean		Global
Squash	*Cucurbita,* et al.	Global
Sugarcane	Vegetative	Global
Sugarcane	Seed	Global
Sweet potato	Seed	Global
Tomato		Global
Wheat	Cultivated	Global
Citrus	Vegetative	N. America
Maize		New World
Rice		Regional

Source: IBPGR Annual Report, 1987, p.29–35

respectively) and scores third for cotton (8 percent). The United States ranks first in sorghum, wheat and barley, second in rice, third in soybeans, fourth in maize, and sixth in potatoes.[40]

In an impressive work by the Prescott-Allens,[41] data are given for 226 crops grown in or imported into the U.S. each year. The farm gate or import value of each crop exceeds one million dollars. Table 5 shows the fifteen crops annually worth more than one billion dollars each to the American economy.

Combined, the 226 produced or imported crops have an economic

Table 5
The Billion-Dollar Crops
U.S. Crop Germplasm Security and the IBPGR Network

Crop Rank[42]	U.S. Farm Gate or U.S. Import Value Sales/ Imports Storage (U.S. $ Millions)	Center of Diversity	U.S. Storage Mandate from IBPGR[43]
Soybean	11,278.4	Chinese	Global
Maizc*	10,412.4	Mesoamerican	Regional
Wheat	6,475.1	Near Eastern	Global
Cotton	4,233.0	African/Andean	Greece
Coffee	3,925.3	African	Import
Tobacco	2,851.4	Andean	Greece
Sugarcane	1,722.5	Southeast Asian	Global
Grape**	1,524.9	Central Asian/Mediterranean	—
Potato	1,206.0	Andean	CGIAR
Rice	1,163.1	Indo-Burma	Regional
Sweet Orange+	1,150.3	Southeast Asian	Regional
Sorghum	1,146.5	African	Global
Alfalfa**	1,053.7	Central Asian/Euro-Siberian	—
Tomato	1,051.0	Andean	Global
Cacao	1,016.0	Andean	Import
	$50,209.6		

*Many authorities consider the U.S. collection to be the largest and most diverse. Although the USSR and Yugoslavia both claim to have extremely large collections, some scientists believe these are no longer viable.
**As of 1987, IBPGR has no network base for this crop.
+IBPGR information is for citrus.
—No information available.

value of over sixty-five billion dollars. The top fifteen crops account for seventy-seven percent (or more than fifty billion dollars) of the total. IBPGR has granted the United States a global mandate for germplasm storage for five of the fifteen top crops, and assigned it regional responsibility for another three. Of the remainder, three are mandated to others, two have no IBPGR mandate at all, and two are only imported. For eleven of the fifteen crops, the United States ranks among the top four holders of stored germplasm in the world. With the assistance of IBPGR, one might say that U.S. food security is improving.

The bias becoming evident has not only aroused the ire of the South. Dr. David Wood of International Center for Tropical Agriculture (CIAT) has also joined the clamor of protest. First in Ethiopia in 1986 at a conference we attended on genetic diversity—where he announced that CIAT would not be part of the IBPGR "network"—and then in an open letter to fellow IARC gene bank directors, Wood has called for a re-evaluation of the network approach. "There is a relation between the amount a country donates *to* IBPGR and the number of collections designated to that country *by* IBPGR," Wood claimed in 1987.[44]

IBPGR's biases do not end with storage networks. It has also been accused of a funding bias toward the North. Though IBPGR is intended to be a "catalyst" encouraging conservation in developing countries, fully fifty-seven percent of all grant monies went to scientists and their institutes in industrialized countries, while another ten percent went to fellow CGIAR institutes. Less than a third of all grants accrued to the South. Some Northern countries, including Austria and France, actually received more grant money from IBPGR than they contributed to it. The British got back eighty percent of their investment and the Americans had two-thirds of their funds returned. For the thirteen industrialized countries contributing to IBPGR's budget in the mid-1980s, the average "kick-back" was thirty-nine percent.

IBPGR has made itself liable for bias accusations in another way. While participating in a major germplasm program with the nine SADCC states of southern Africa, IBPGR has also contemplated some quiet work with South Africa. According to board minutes, the executive committee sought a "clarification of [the] CGIAR position on links with South Africa . . ." The minutes go on to report that "the Chairman of CGIAR suggests that IBPGR, if the work is considered to be of importance, could carry out a low key programme in that country."[45]

This says as much or more about CGIAR as it does about IBPGR. The CGIAR chair's reply emphasizes that the dispute over IBPGR must shift to the whole of the Consultative Group on International Agricultural Research. CGIAR's defense of IBPGR has been tepid, at best, but it has nonetheless insulated one of its member institutions from the slings and arrows of outraged democracy at FAO.

How did this happen? The IBPGR board is, effectively, self-elected. In the years since its founding in 1974, the ratio of board members from North and South has been more or less reasonable, considering the technical nature of its work: a few more in the North. But those from the North tend to stay with the board much longer, an average of six years compared to an average of three for Southern scientists. In short, the board is an "old boys club" largely made up of British and American colleagues who tend to finance the scientists and institutes they know best. In a very real sense, IBPGR was a flag of convenience under which developed countries, cloaked in the legitimacy of the UN, could collect seed in the South and store it close to their breeders in the North. Their cash contributions to IBPGR for conservation in the South acted as a cheap subsidy for their own national collection priorities. Instead of an average collection cost of forty-two dollars per sample, industrialized countries used IBPGR to gather seed at twenty-six dollars per sample.[46]

All along the way, the irascible opponent of Third World countries' influence over their own germplasm donations has been the United States. Backed by satellite governments such as Canada, the U.S. has argued vehemently that the exchange of genetic resources was not a problem. At every opportunity, U.S. officials and CGIAR staffers pointed to embargoes of plantation crop germplasm by Third World governments, or tardiness by Eastern European states in passing on samples.

But the reality is more complex. Germplasm embargoes are common. Most of them—as the North contends—are instigated by the South. Ethiopia does embargo coffee germplasm.[47] India embargoes black pepper.[48] According to internal IBPGR field reports, Thailand, Indonesia and Malaysia make the exchange of tropical fruit germplasm very difficult. Tropical vegetables and sugarcane are often subject to unofficial embargoes also. Although Brazil is prepared to negotiate access to rubber germplasm, free exchange is by no means automatic.[49] Ecuador is less than forthcoming with cocoa and Turkey takes a hard line on tobacco.[50]

The foregoing are hardly food staples, however. And some of the restrictions credited to the South are actually based on the policies of private companies located in the North. For example, Liberia's rubber embargo is traceable to the Firestone Tire Company[51]—not to the Liberian government. Corporate constraints are also behind the Honduras restrictions on banana germplasm, which was controlled by United Brands until the company jettisoned its collection in 1983. The problems of exchange of tea clones in Kenya and sugarcane in Hawaii resulted from corporate barriers. Oil palm germplasm is the private preserve of Unilever.[52] In general, the germplasm restrictions of most Third World states are nothing more than the continuation of policies observed by foreign plantation enterprises during the colonial era. Rarely do Third World embargoes involve basic food crops.

The serious political restrictions on germplasm exchange come from the North. National plant breeders' rights legislation is a major factor. Sugar-beet breeders in Europe routinely impose embargoes on the exchange of breeding material given to public gene banks for storage.[53] Individual companies such as Hurst, Gunson in the United Kingdom, and Cebeco and Nunhems Zaden in Holland all admit to embargoes.[54] On the other side of the Atlantic, Campbell Soup imposes restrictions on tomatoes.[55] More serious is the restriction imposed by the Dublin gene bank, which will not supply material to scientists from any country not a member of UPOV (International Union for the Protection of New Varieties of Plants—the plant patenting agency in Geneva).

The whole debate at FAO can be likened to a domestic squabble in which one partner claims there is a problem and the other claims there is no problem. The only thing neighbors can be sure of is that there is a problem. Our own research on the issue of germplasm exchange tells us that, in principle, the North has no desire or intention to embargo germplasm to other countries. We are convinced that the overwhelming majority of Northern scientists and gene banks—in the public sector—see real advantages to the full and free exchange of all genes. And when embargoes are imposed, scientists often find a way around them.

However, since whether germplasm is "raw" or "improved" tends to depend on the view of the holder, it is simply impossible for industrialized countries to guarantee that commercially held germplasm will be freely exchanged. This also means that materials stored in public banks can be subject to various forms of embargo in order to safeguard the perceived market interests of companies. While espousing free exchange, Northern governments are blind to those legislative and commercial pressures that dampen their own full exchange of germplasm.

Although they would be the last to remember, Northern governments and scientists have also participated in their own embargoes. Consider, for example, the Russian grain embargo in 1979–80. Related to the Soviet invasion of Afghanistan, the U.S.-imposed boycott unavoidably included germplasm, despite U.S. policy concerning free exchange. Canadian government officials—pressured into the embargo by the U.S.—have admitted that germplasm was caught up in the same sweeping government directive and, for about six months, gene bank officials were prevented from exchanging seed with the Soviet Union.[56]

That the United States does impose political constraints to the exchange of germplasm has been fiercely debated throughout the 1980s. Much of the debate in Rome over the Undertaking was fueled by the circulation of a 1977 letter from the United States Department of Agriculture (USDA) to IBPGR.[57] The letter frankly admitted that germplasm received from the international community would become the property of the United States and that such seeds would be subject to fluctuations in American foreign policy.

Prior to FAO's 1981 conference, we talked on the telephone to Dr. Quentin Jones, then head of the USDA National Plant Germplasm System and an IBPGR Board member. Jones confirmed that the policy stated in the 1977 letter was still intact. Asked for a list of countries on the U.S. "blacklist," he denied any knowledge of such a list, but added that embargoes either had been or were in effect against a number of countries. These included Afghanistan, Albania, Cuba, Iran, Libya, and Nicaragua as well as the Soviet Union.

Yet, a few months later, Dr. Jones, in reply to a letter from a Cornell University professor,[58] defended the U.S. germplasm exchange policy, arguing that only technical diplomatic problems occasionally required routing germplasm through a third country and citing the example of Albania.[59]

A cable from the U.S. Embassy in Rome to the State Department in Washington in September, 1984, shows that, contrary to other comments, IBPGR was well aware of past U.S. germplasm embargoes:

> Williams [Trevor Williams of IBPGR] also commented that while he understands U.S. has never formally denied request for germplasm, there have been instances when the word went out that "for diplomatic reasons" answers to germplasm requests should be slowed down, not acted on, or answered through intermediaries (e.g. Canada). Williams understood this was case with Cuba after crises of early 1960s and with Angola in mid-1970s.[60]

If all this were news to the State Department, it was obviously well-known to the international scientific community. A telex from the American Embassy in The Hague a month later confirmed the views expressed by Trevor Williams.[61] The telex quotes a letter from Jaop Hardon, acting director of the Dutch gene bank and a longtime representative for the Netherlands in Rome, saying:

> Finally, you state that the U.S. is committed to the principle of free and unrestricted exchange of germplasm with all nations. I trust that this means that previous restrictions imposed on such countries as Cuba, Nicaragua, and Vietnam have recently been lifted.

Despite all this—in Rome and back home—the United States government continued to deny that embargoes were a reality. Internally, however, the Department of Agriculture was feeling the heat. Jones' predecessor, Henry Shands, widely credited with bringing a more open and conciliatory style to the U.S. germplasm program, favored a clear policy endorsing free and full exchange. And the USDA's Paul Fitzgerald was dispatched to negotiate with the Treasury Department officials that govern trade embargoes. On July 25, 1986, Fitzgerald reported: "I had a pleasant telephone call and a most welcomed message from Cheryl Opacinch, Office of Foreign Assets Control, Treasury Department, . . . regarding my earlier request for exemption of plant germplasm from the embargoes imposed by the U.S. on North Korea, Cuba, Viet Nam, Cambodia, Nicaragua, and Libya."[62]

When the problem was still not resolved, the USDA sent out a "fact sheet" to key players on the embargo issue in the Reagan Administration.[63] Dated February 12, 1987, this memorandum outlined the problem and proposed a meeting to try to clear up the policy confusion. According to the memo, material from South Africa and another unidentified country could not be received, and material for Nicaragua could not be sent. While emphasizing the confusion, the memo also explicitly concedes that the United States does embargo germplasm to some countries for political reasons. The title says it all: "Subject: Movement of Germplasm to and from Embargoed Nations." Given the intensity of the debate at FAO, passages from the memo are worth quoting:

> [3.] Present situation:
> a. Three amaranths destined for Nicaragua (from Rodale) have been awaiting approval since November.
> b. A request to import virus-free fruit stock from South Africa was awaiting certification of the originator by APHIS when the commercial embargo (in and out) on S.A. was put in effect.

c. A request for a large quantity of grass seed by an ARS scientist in Arizona for arid range studies has since been caught in the embargo issue.

4. Sensitivities:

a. FAO activists will denounce 3.a. as proving the U.S. denies germplasm requests.

Other points:

a. The real question seems to be: Does the exchange of germplasm override foreign policy? (i.e. embargo) State seems very uncomfortable to be put on this spot. This may be a "put up or shut up" situation regarding germplasm.

b. It may take Congressional declaration of germplasm being above policy to get future action.

c. Declaration of germplasm as a humanitarian aid item such as medicinal supplies still makes it subject to export controls.

In September, 1988, we contacted Rodale (the publisher of *Organic Gardening*) to determine the status of the amaranth material and were informed that Dr. George White, plant introduction officer for USDA, had advised Rodale simply to mail the seeds to Nicaragua first class.[64] Sources within USDA confirmed that formal approval was never given. Rodale had no confirmation of the arrival of the seed in Nicaragua and arranged to ship duplicate samples to us in Canada for transshipment to Nicaragua.

The world cannot afford political ambiguity from its most important germplasm holder. American reluctance to participate in the FAO Commission and Undertaking on Plant Genetic Resources and to support the formation of a fund gives other countries justifiable cause for concern. Certainly, if one of the world's most open states (in terms of germplasm exchange) accepts political embargoes, there is need for careful intergovernmental oversight, through the United Nations, of all germplasm exchange policies.

For a while, it seemed that key members of the Reagan Administration agreed. In July, 1984, U.S. Secretary of State George Shultz sent the following telex to the American Embassy in Rome:

Subject: Plant Genetic Resources Commission

Please inform the FAO that, in response to the invitation of April 6, 1984, the U.S. wishes to be considered a member of the Commission on Plant Genetic Resources. Comment: In joining the Commission

on Plant Genetic Resources the U.S. does not indicate approval of the FAO Undertaking on Plant Genetic Resources.
Shultz

The message was either countermanded or never received. Two forces came together to halt the "normalization" of U.S. foreign policy on this issue. First and foremost, the American seed industry—which has attended FAO Commission meetings as part of the U.S. government's own "observer" delegation, blocked the policy change. Secondly, in a related move, officials inside USDA recognized that the long-term impact of the Shultz cable would be to bring IBPGR to heel under the United Nations flag.

In the absence of clearly defined international agreements, political problems are inevitable. Ambiguity—in situations where one side has power and the other side does not—tends to serve the interests of the powerful. In the case of genetic resources, the benefits are clear. The South has the raw germplasm in its forests and fields, the North has much of the South's stored germplasm in its gene banks. As long as the material continues to flow North, industrialized nations have no reason to bother much with developing intergovernmental structures and protocols.

In May, 1986, the seeds issue surfaced at the American Association for the Advancement of Science meeting in Philadelphia. On the week-long agenda was the title: "Seeds and Sovereignty: Debate Over Control of Plant Genetic Resources." The combatants were M.S. Swaminathan of IRRI, William Brown of Pioneer Hi-Bred and Jack Kloppenburg of the University of Wisconsin. After years of backroom discussion, the U.S. scientific community was finally recognizing that the North-South split over control of germplasm would not just go away.

For the crowd gathered at the Philadelphia Holiday Inn, however, the issue was not rubber or cinchona, but food. Kloppenburg had extrapolated from FAO data on crop production and linked the figures to a modified version of the Vavilov centers to describe the interdependence of all the centers for future food and industrial crop germplasm requirements. In one sense, the somewhat arbitrary calculations could give encouragement to those who would argue that each Vavilov center sorely needs the others. Even the most genetically self-sufficient ones, such as west central Asia, look to other centers for almost a third of their basic food germplasm. But, on a more realistic continental basis, Latin America

Table 6
Food Crop Independence
Percentage of Major Food Crops from within Selected Regions or Continents of Origin

Region	Percentage Self-Reliant
North	6.4
Africa	12.3
Latin America	52.2
Asia	56.8

and Asia are the clear heavyweights, never accounting for less than three-quarters of any continent's gene needs. The North (Australia, Europe and North America together) account for less than seven percent of their own needs and never contribute even a full percentage point to the South.[65]

Nevertheless, the North has used Kloppenburg's figures in a "divide and rule" strategy to insist that some Third World regions would be disadvantaged by a "centers of origin" approach to assigning benefits to those who "donate" genetic resources. They claim that even the less well-endowed Third World regions benefit from the current system. But that interpretation of the Kloppenburg table does not reflect either political or practical realities.

In the first place, unlike their counterparts in the industrialized countries, poor farmers in developing countries are far less dependent upon exotic germplasm since they are surrounded by much greater variability. They generally look to a wider range of crops (teff in Ethiopia, African rice in Senegal) than are presented in the Kloppenburg table (twenty food crops only).

Secondly, genetic diversity in the centers given for Europe (the Mediterranean and Euro-Siberian centers) is the most eroded in the world. Moreover, a noteworthy part of the Mediterranean center lies within Africa (consider wheat and important forage crops), but Africa is rarely "credited" for its contributions to Mediterranean crops.

The four centers defined by Kloppenburg for Asia share several crops among themselves and are ancient centers of production for many of the crops originating in other Asian centers with access to considerable diversity. Many of the Middle East cereals, for example, have a long history in

India and China. Rice has spread throughout Southern Asia across at least three centers.

For all the reasons cited above, the political "pain threshold" for Australia, Europe, and North America—with their highly uniform plant varieties and mechanized food processing—is much lower than the threshold for Africa, Asia, or Latin America.

Table 6 restructures the Kloppenburg study on the basis of the North and Africa, Asia and Latin America.

The great dependence of the North is clear from the table. However, Africa does not appear to fare much better. This, in part, is because of the selection of the twenty crops (Ethiopia's premier crop, teff, is not included, for example), and is partly a reflection of reality. But, even for Africa, the "pain threshold" is much higher than indicated since many exotic crops such as wheat and barley have a long history in Africa and, again, most farmers work with a much wider range of crop diversity than do their Northern counterparts.

Biopolitics. From the earliest times, ownership and control of plants and their diversity have been much more than merely scientific or technical concerns. They have been and will continue to be profoundly political. The strength of nations has risen and fallen; great fortunes have been made and lost; and people have enjoyed plenty or suffered hunger at least in part because of who owned, controlled, used, and benefitted from genetic diversity, and who did not.

Too often the story is told from the perspective of the scientist, plant breeder, company, or industrialized country—those who have so much and yet need even more. It is tempting to view their needs and theirs alone as legitimate. It is tempting to argue that any effective or efficient conservation system must be designed to serve them. In this chapter and in the next we have tried to balance the legitimate needs of industry with the very real needs of the farmers themselves for genetic diversity. The logical next step is to look at what is being done, and what could be done, to conserve genetic diversity in a manner which benefits not just some people, but everyone.

CHAPTER TEN

Responsibility and Commitment

It must be taken as a cardinal responsibility that we do not destroy what we cannot re-create and do not yet comprehend.

—R. J. Goodland and H. S. Irwin

Most of us want our crises in single file. Where does this one rank among the legions of disasters awaiting the human species? What's going to get us first? Loss of genetic diversity? Nuclear war? The greenhouse effect? Breakdown of the ozone layer? Television? Acid rain? Soil erosion?

Those who care about the future of people and the planet do not have to play issue-oriented Russian roulette. All the issues come down to questions of social justice and public action. Working on one does not mean losing sight of the others. We do not want this book to steal anybody away from their work on nuclear disarmament or industrial pollution to grapple solely with seeds.

But neither do we want to underplay the importance we place on the conservation and utilization of genetic resources. Genetic erosion—not a new problem—threatens to intensify as biotechnology produces new and more attractive varieties for farmers to adopt. Our concern for genetic diversity and our action to conserve it must not diminish. But we need to make connections between this issue and others. One very troubling example of the need to understand the interrelationships will suffice.

In previous chapters we have spoken of the need to conserve diversity because of its importance to evolution, to plant breeding, ecologically sound agriculture, human nutrition, culture, and to peoples' attempts to become self-reliant. In the section below we briefly examine the connection between genetic diversity and the military.

The Romans had the right perspective. They respected not only Mars, the god of war, but also Robigus, the god of wheat rusts. Both, they knew, were killers. If we have to make an insane bet, then we must assume that the world has more to fear from nuclear holocaust than wheat rust. But, like the Romans, we want to hedge our bets, for the age of biotechnology is uniting Mars and Robigus to make war with hunger and disease.

Rome was no stranger to biological warfare, as ancient written condemnations of well poisoning confirm. By the Middle Ages, the Mongols were helping spread the plague in Europe by placing infected bodies on their hurling machines to be pitched over the walls of cities they were attacking. Hundreds of years later the British gave blankets infected with smallpox to tribes of American Indians to "extirpate this execrable race." Poison gas was used with ugly effect by Germany in World War I.[1]

By the Second World War, every major power was prepared with germs and chemicals. The British tested anthrax on a Scottish island, which remains uninhabitable to this day. The Americans stockpiled an immense amount of poison gas and constructed a facility in Indiana to mass produce deadly diseases.[2] Japan, however, went one step further.

Japan established a huge research and testing facility near Pingfan in Manchuria. With a staff of three thousand they worked on some of the world's most deadly diseases: anthrax, cholera, typhoid, plague, glanders, tsutsugamushi, smallpox and tuberculosis. Plant diseases were also investigated. The records of these experiments, obtained after the war by the U.S. in exchange for grants of immunity from war crimes prosecution, revealed the sacrifice of hundreds of sheep, horses, mice and guinea pigs.

More horrifying, the records contained numerous references to experiments with monkeys. Years later researchers were able to confirm that the "monkeys" were in fact human beings—Russian, Chinese, Korean, British, Australian, and American prisoners of war—at least three thousand of them.[3]

More recently, abundant though circumstantial evidence exists of

American use of biological warfare (anthrax, cholera, plague and smallpox) in the Korean War. And charges continue to be leveled at the U.S. over the first appearance of the deadly African swine fever in the western hemisphere, in Cuba in 1971. About the same time, in testimony before a Senate committee on intelligence activities, Central Intelligence Agency (CIA) officials acknowledged that they had developed "methods and systems for carrying out a covert attack against crops and causing severe crop loss."[4] After World War II, the U.S. also engaged in a massive program to test dispersal of germ warfare agents inside the U.S., using its own population as guinea pigs. Supposedly harmless bacteria were sprayed in the New York City subway system, in Washington's National Airport, all over San Francisco, in midwestern stockyards and university poultry farms, and in dozens of other locations—including Winnipeg, Canada.

Despite this history, the use of biological weapons remained problematic until recently. Effective use implied the existence of a defense to protect the aggressor against the effects of its own weapons. Precise targeting was difficult. Spreading diseases has a way of coming back to haunt the offender. Moreover, biological warfare was and is considered dirty. Countries proud of bombing the daylights out of an enemy cloaked their biological and chemical warfare activities in secrecy. Finally, legal conventions outlawing biological warfare have been signed by most nations.

But the advent of biotechnology gave new hope to germ warriors. With biotechnology, naturally infectious agents can be increased in virulence, harmful genes can be inserted into benign organisms, and the new creations can be made resistant to known antibodies and vaccines. Biotechnology makes possible sophisticated targeting. And it even solves the problem of the ancient social taboo. Who is to say if the new disease is a natural mutation of an old disease or the creation of military scientists? Under the microscope, there is no way to tell.

In the U.S., President Ronald Reagan reversed the course set by three previous administrations and began a massive buildup of chemical and biological weapons. During the Reagan years the budget for the chemical and biological warfare program grew to a billion dollars annually. The number of unclassified military projects using recombinant DNA and monoclonal antibodies rose from zero in 1980 to over a hundred in Reagan's first term. Meanwhile, federal support for nonmilitary university research in the life sciences declined.

While exotic biological weapons for use on humans grab the occasional headline (for example, the revelation that the United States and the USSR were both probably trying to insert genes coding for cobra venom into influenza), the most practical and likely use of biological warfare involves agriculture.

Biological warfare directed towards agriculture arouses less suspicion. It does not necessitate the vaccination of one's own population against an exotic disease. Used in the Third World in situations of "low-intensity" warfare, it can be quite effective. Destruction of an enemy's coffee crop might accomplish more with less cost and risk than sending in the marines. And it would provide "plausible deniability," because the development, production, stockpiling, and dissemination of plant diseases can be accomplished with little risk of detection. In fact, were such warfare being waged today, we would probably not know it.

Because they are undetectable, biological weapons are also unpoliceable. There is absolutely no way that any government can be assured that another government does not have hidden stockpiles. The CIA kept its little stockpile of diseases despite a 1969 presidential order that it be destroyed.

Through biotechnology, genetic resources become both the seeds of life and the seeds of destruction. As governments spend billions seeking out and developing new and more lethal diseases, others concentrate on preserving disease and pest resistance in the form of genetic diversity. The evil that biotechnology makes possible provides yet another reason to be concerned about the control and conservation of genetic resources—the preservation of our options for the future. Can we afford to throw away the diversity that might help us rebuild agriculture from the destruction of biological warfare? Can we risk losing the diversity of life in this age of biotechnology?

COMMITMENT IN ETHIOPIA

Back in the early 1970s, a scientist from Purdue University made the journey to Ethiopia to gather sorghum. Some years later, we were told that he forwarded a copy of his laboratory analysis of the farmer cultivars to his Ethiopian hosts. According to the report, he had "discovered" that one sorghum accession had a very high protein content and excellent baking qualities. He could have saved himself some laboratory time had it occurred to him to ask the farmer who first gave him the seed about its characteristics. Ethiopians call their variety *sinde lemine*, which translates as "why bother with wheat?"

When Yilma Kebede tells the story, he literally shakes with laughter. Lounging with one leg stretched out on an office sofa, Dr. Yilma talks of another high-lysine sorghum, the local name for which is "milk in my cheeks."[5] As team leader for sorghum breeding at the research institute in Nazret, Yilma has developed a healthy respect for Ethiopian farmers and their contribution to sorghum. His natural easy-going style left him though, when he recalled an earlier visit from Ciba-Geigy officials who tried to sell hybrid grain sorghum to his government. "It is ours," he told us. "Sorghum originated here in Ethiopia."[6]

Across the room, Yilma's colleague, Dr. Melaku Worede, shares both his irritation and his solutions. Melaku is charged with one of the toughest and most important jobs in the world. He is the director of the Plant Genetic Resources Center, the genetic conservation campaign for Ethiopia.

In and of itself, Ethiopia could be regarded as a Vavilov center. Its fantastic terrain of mountains, valleys and plateaus, combined with a long history of cultivation, make the country one of the most botanically diverse and important points on the globe. Ethiopia is home to major world crops like sorghum and many millets, as well as coffee. Less well-known outside the country is its teff crop, which is still the most important food staple. Thousands of years of farming have made the region a secondary center of diversity for wheat and barley as well.

But Melaku Worede stresses that his country's ragged landscape is only part of the story. The other part is its people. "A farmer will take me to his bin and I will look in at the barley or teff or sorghum and I will see nothing. To me, it looks the same." Melaku waves his arm. "But the farmer will just reach in and show me that this one is for this soil and this one is for that and this one makes good injura [Ethiopian bread made with fermented dough] and so on. I am the scientist with the training. But the farmer knows his seed."

With a sense of humility, Melaku Worede and his highly trained team are working closely with Ethiopian farmers to conserve crop genetic diversity. They are painfully aware that laboratory studies of yield and nutrition cannot reveal the true value of an accession.

When Ethiopian plant explorers go collecting, they take along questionnaires to be filled out with farmers for each accession. In effect, much of the documentation and evaluation work can be done with the farmer right in the field. During the drought, farmers became more involved and helped directly in collecting in two of the worst hit regions.[7]

While the Ethiopian gene bank was originally built by a West German agency, Ethiopians are in full control. Staff have been schooled in Sweden, the Soviet Union, Canada, and the United States. Seed germination levels are closely monitored, as are temperature and humidity controls. Seeds are hand-cleaned and inspected before storage. Everything is checked, double-checked, and dutifully recorded on a Hewlett-Packard computer. If there is any fluctuation in power supply, the alarm bells ring not only at the bank but also in Melaku's bedroom at home.

Yet the Ethiopians are convinced that even these precautions are not enough. Together with the Ethiopian Seed Corporation, Melaku is working to establish as many as twelve satellite seed storage units that would allow farmers' seed for each region to be kept in that region. The "banks" would not be glamorous, but they would be protection against future droughts. In such times farmers could get locally adapted seed from these units. In the past, farmers have had to eat the seed that in normal times would have been saved to plant the next crop. This has forced many farmers to resort to aid shipments and imported seed for planting, with often disastrous results: the crop planted with the imported seed fails, and the old locally adapted seed has disappeared—forever. Melaku knows that if traditional seeds, adapted to the Ethiopian environment, are not saved, Ethiopia will never be able to build a self-reliant agricultural system. The famine will become permanent. A request to the World Bank for funding the satellite units was turned down.[8] Some support from Canadian non-government and church agencies has come forward, but more is needed.

The international community has never given Ethiopia's traditional seeds or crops their proper due. Statistics available on crop harvests in the country from the late 1940s omit any reference to Ethiopia's dominant food crop, teff, until the early 1970s. From nowhere in 1948, teff suddenly occupied over twenty-seven percent of the cropland in 1972. Half a dozen other traditional crops—all important to local diets—have also been ignored.

Aside from teff's meteoric rise on the charts, the big changes in Ethiopian agriculture are in corn, wheat and oats. Wheat is no stranger to the country, but by the end of the 1970s, thirty-seven percent of the wheat land was sown to "improved" or "high-response" cultivars.[9] Where hundreds if not thousands of distinct varieties once grew, four now dominate the landscape.[10] Oats production did not become a statistical factor in

Ethiopia until the mid-1970s and new cultivars are now replacing more traditional crops in order to provide livestock feed.

But the real up-and-coming crop is corn. The Portuguese likely brought corn to Ethiopia centuries ago, but the crop was not significant until the early days of the green revolution. Between 1972 and 1979 alone, corn grew from about ten percent to almost eighteen percent of the national harvest. Today, one variety is dominant.[11]

As crops like corn and oats and new cultivars of wheat spread, old crops like teff, barley, and even sorghum have gone into decline. The fate of many poor people's crops is unrecorded. Yet, in 1985, when Jan Engels—a Dutch agronomist attached to the Ethiopian gene bank—was on a trip with his family, he passed mile after mile of what he first thought were onions. Only when he stopped his car and got out to inspect the plants closely did he realize that he was looking at the death of the corn crop.[12] The high-response wheats suffered the same fate: they died of thirst.

The Ethiopian famine had many causes—overgrazing, water management problems, politics, the drought itself. But unnoticed among those problems was the pressure imposed by outside "experts" for Ethiopia to abandon its drought-tolerant crops and cultivars in favor of green revolution varieties. The old seeds may have been low-yielding, but they didn't need much water, they would germinate even after long periods of drought, and there was always something to harvest at the end of the season. Teff may or may not be the most nutritious crop in the world (the research is so limited that nobody knows for sure) but it must rank as one of the toughest. In 1985, Ethiopia's bilateral government aid donors were still talking (at least among themselves) of pressuring the government to abandon its teff seed for modern high-yielding corn.[13]

Throughout the famine the Plant Genetic Resources Center dispatched scientists in jeeps and on donkeys almost every day to search the fields, bins, and hills for traditional seeds that might otherwise have become extinct. Melaku Worede and his allies at the Ethiopian Seed Corporation recognize that the food security of Ethiopia may depend upon the survival of the old landraces.

SOWING REVOLUTION

Under similarly stressful conditions a continent away, Nicaraguans are learning the same lesson. The welfare of the peasant farmers and the survival of the nation depend on creation of a self-reliant, sustainable system of agriculture. And such a system of agriculture depends on seeds adapted to the environments and social conditions of the country.

Once Nicaragua was the seed distribution point for Pioneer Hi-Bred's Central American operations. But the company pulled up stakes after the 1979 revolution, and took with it most of the trained personnel needed to clean and grade seeds and keep machinery running. Faced with a facility too complex and sophisticated for local needs, the Sandanistas set to work learning everything they needed to know about seed processing and marketing. The result is Emprasem, an unattractive collection of high wooden sheds strung together like boxcars on the outskirts of Managua. Aside from some shiny new Danish seed dryers (donated by the Canadian Save the Children Fund) and some other equipment given by the Canadian University Service Overseas (a volunteer service agency), the machinery churning up dust in the great sheds is being kept together, literally, with prayers and wire. There are never-ending problems with seed grading screens, dust, and spare parts.

There is also a problem with seed. An old hand in oily jeans and a blue denim shirt joked with us in Spanish about the corn lines bought the year before from an international company. The company forgot to warn the Nicaraguans to stagger the planting times. When the female line was ready, the males hadn't reached puberty. The old man's smile didn't reach as far as his eyes. The company had charged fifteen cents a kernel.

A few days later, up north in Esteli, the twenty-one-year-old director of the regional agricultural research station wouldn't let us leave until he had rummaged through the warehouse, pulling down boxes of imported seed for us to examine. Old seed, chemically coated seed, broken seed, and an American garlic variety that had utterly failed to germinate. Good statistics are scarce, but these seed imports are costing Nicaragua a small fortune in foreign exchange. Until the embargo in 1985, Nicaragua was one of the largest buyers of American seed. The previous year they paid $648,000 for sorghum seed alone.

During our weeks in Nicaragua in 1984 we were taken to several other stations and to the government seed procurement desks located behind the agricultural university in Managua. We saw catalogues and packets from Desert Seed (Atlantic Richfield), Petoseed (Ball), Ohelsons Enke (Swalof) and Northrup King (Sandoz). Hunched with frustration over his catalogues, the young man responsible for horticultural seed imports told us his main sources were IPB, Asgrow and Ferry-Morse. At Sebaco, we saw

Sandanista posters alongside brightly colored counterparts from IPB (Shell) and Asgrow (Upjohn).

Humberto Tapia, then the country's director of plant breeding and imports, wants to do something about it. A tall, stern-looking administrator, Tapia is also a poet and plant collector, and is highly regarded as a scientist throughout Central America. His aloof shyness falls away when he talks about the diversity of corn and beans he has found in the countryside.

One evening, Diane Palaez (a Canadian CUSO cooperant) and Aurelio Ilano, the country's only bean breeders, dumped out a large bag of seed to show us. Even untutored eyes could see that each of the hundreds of beans pouring over Aurelio's desk was different. Sifting through the pastel rainbow, Palaez and Ilano talked enthusiastically about the cooking properties of one ("fast cooking means less fuel") or the disease-resistance of another ("nothing seems to touch it"). At Palaez's urging, Ilano pulled from his center drawer the graphs and tables he had presented to a Central American conference in Panama in 1983. The tables compared the trial results of CIAT's green revolution bean varieties to Nicaragua's own adapted landraces. The trials showed conclusively that CIAT's varieties substantially outperformed the local champions under low-stress, high-input conditions. Under any other conditions, CIAT beans proved erratic performers and Nicaragua's beans led.

With a wide grin, Aurelio Ilano leaned across his desk. "Did you know that Nicaragua is exporting revolution?" Not waiting for a reply, the scientist opened his hand to display three beans: "Revolution seventy-nine, Revolution eighty-two and Revolution eighty-three." He laughed happily. "These are local varieties we have adapted. Now we are exporting them to Costa Rica, Honduras, and even El Salvador. "Except there," his grin widened, "they are called R-79, R-82, and R-83!"

The Nicaraguans are also exporting cotton varieties and some of their open-pollinated corn holds great promise. Not all the work began with the revolution, but Tapia and his staff now have the energetic support of Jaime Wheelock, Nicaragua's hard-driving minister of agriculture and agrarian reform.

Like just about everybody else we met in Nicaragua, Wheelock is young. He also shares the national penchant for self-deprecating humor. "I'm a lawyer and a social historian," he told us, "not a *campesino*. When I took on this job I didn't even know how to plant corn." His eyes took on a rueful twinkle, "I still don't!"

But Wheelock has won the respect of the campesinos (peasant farmers) and the international community for his successful land reform program and his support for new cooperative farms. He is also the first to tell the negative stories—the hand-spraying of one of the world's most hazardous chemicals in the cotton fields, and the regional director who mistakenly interplanted cotton and corn. He is firm on a number of points: Nicaragua cannot depend on foreign seed imports because they are too costly and politically unreliable. To safeguard against natural or induced plant diseases, campesinos need access to diverse, high quality local varieties. Finally, Wheelock asserts that Nicaragua has an abundance of plant genetic diversity which must be exploited.

The achievements of Humberto Tapia and his small staff of scientists are seen by Wheelock as the direction Nicaraguan agriculture has to take. The missing link has been an organized campaign to collect and evaluate Nicaragua's genetic resources. In 1984, that task fell to Francisco Berrios, vice-minister and a key government strategist, and the ministry's then director of research, Bayardo Serrano.

Berrios and Serrano are giving top priority to a national genetic resources program which is to gather at least nine thousand accessions and will feed at least four useful new accessions to Tapia and others for breeding within two years. The plan calls for one long-term gene bank plus three medium-term banks. Work is already well under way on another medium-term bank and still another is on the drawing boards in Esteli.

By the end of 1984, Wheelock's team had assembled four agronomists responsible for plant collecting and evaluation. In addition, the government has turned over to scientists an island for animal breeding and for seed rejuvenation. Still more land is ready close to Managua. With other ministries, Berrios is searching for locations for biosphere reserves.

These steps alone give Nicaragua a conservation strategy better thought out than any for the Amazon. Daniel Querol, the young, intense geneticist asked by Berrios to launch the campaign, has gone further. After participating in collection and documentation work from Peru to the Philippines, Querol sees the value of involving campesinos. A key part of the new strategy is a system of campesino/curators—peasant farmers and cooperatives asked to conserve endangered landraces on their own land. To emphasize the importance of these varieties and to compensate participants for any lost productivity, Querol offers to pay farmers about nine dollars a landrace. So far, most farmers have refused any money.

This concept is linked to another innovation: a campaign to make Nicaraguans proud of their plant genetic diversity and to encourage consumers and gardeners to look for a wide assortment of varieties. Unless people use diversity, Querol contends, any institutional effort will be moribund.

In the short years since Nicaragua began its conservation campaign, it has grown to become a model for Latin America. More than half the entire agricultural research budget is now devoted to genetic resources collection and utilization.[14] This is in addition to the important support Nicaragua has received for this work from the Canadian agency Inter Pares, and the Lay Movement for Latin America in Rome. A multidisciplinary team of thirty agronomists and technicians is searching the fields and forests for everything from beans and medicinal plants to old cattle breeds. Despite regional political tensions, Nicaragua has become the acknowledged leader in both policy and practice.

With mercenaries massing across the border, outside observers might at first find it hard to understand Nicaragua's concern for plant diversity. Wheelock and his colleagues have no doubts. As they see it, national security is impossible without food security, and food security is impossible unless the country controls and develops its crop genetic resources.

By mid-1985, the long anticipated U.S. trade embargo was in effect and Nicaragua lost access to its major traditional seed source. Germplasm resources—even those donated for storage by Nicaragua to the United States—were also cut off. Although the embargo is a painful blow, Wheelock and his colleagues are confident that their timely work on germplasm collection and seed production will be their salvation.

The Contras seem to agree. A vegetable seed warehouse was blown to smithereens in Managua one night.[15] Agronomists working on banana germplasm near the Atlantic Coast were attacked and forced to fight for their lives. The trees were not so lucky; a number were burned down. One evening near Esteli, we went with Daniel Querol to check out the seed potato harvest in a new cooperative. The women and men coming in from the fields had bags of seed potatoes slung over one shoulder and carbines over the other.

STAR, MOON, BULLET, AND BALLOT

Some years ago, Stevie Wonder dedicated a record album to seeds and sang "A Seed's a Star." To Dr. Daisy Dharmaraj, director of

PREPARE, a rural reconstruction and disaster response service operating out of Madras in southern India, the seed is a moon. A series of batiks by a local artist depicts the moon and asks, "Who owns the moon?" The images move on to show a seed and the question, "Who owns the seed?" The batiks are part of a village-based drive to collect and conserve traditional crop varieties at the community level. Farmers and local organizations see not only the need to save seeds, but also the importance of saving them in the community, in context. Socially and technologically, the national gene bank at New Delhi is too far away.

So far the efforts to gather seeds have been haphazard, but the need is recognized and the will is there. When Eva Lachkovics (of Rural Advancement Fund International and the Austrian Institute for International Co-operation program) visited Yhanjavur, a town in Tamil Nadu, she found farmers working with PREPARE to seek out old rice varieties that can now be found only in the most remote villages. Further north at Ahmedabad, Lachkovics found Korah Mathan and the village groups he works with still better organized. Having already held workshops on traditional rice varieties and the problems of high-response seeds, groups there and at Auroraville are well into practical conservation work and linked to a wide range of other seed savers in other parts of India. Again, the government's work is not seen as either reliable or sufficient. Farmers have to take responsibility for their own diversity.

To Didi Soetomo of central Java in Indonesia, the seed is a bullet aimed at the heart of farmers. His organization, Yayasan Sosial Sidomakmur, is working in the villages with farmers. As we walked with him through a village and past rice paddies, he pointed out the modern rice varieties IR-36 and IR-38, and angrily talked about the costs of seeds, fertilizers, and pesticides.

Soetomo and his colleagues argue that traditional rices are often more reliable, more nutritious, and less costly to grow. Farmers cannot afford the green revolution rices but the government has made it flatly illegal to grow the old seeds. The response has been to build a community rice seed bank. Ten varieties were being grown in 1987 and there were more to come. Next they plan a bank for traditional fruits endangered by deforestation and herbicide spraying.

Didi Soetomo, chainsmoking in the midst of the wicker and plastic of his crowded village office, is not your classic nature lover. His concern is the farmers and villagers that surround him. But, as he says, he can't help the farmers without saving their traditional seeds. If they lose control of their seed, they lose control of all agricultural inputs and markets. Genetic conservation is a practical human necessity. When a Yayasan Sosial Sidomakmur representative talks to farmers about their seeds, they understand. The seed is the beginning point for a much wider conversation about the future of Indonesia.

For Rene Salazar, the seed is a ballot. Philippine farmers are voting against the green revolution as they move away from high-response rice and corn toward improved traditional varieties. Thus far the post-Marcos government of Corazon Aquino has failed to improve the lot of small farmers. Salazar, who works with the umbrella group SIBOL NG AGHAM AT AKMANG TEKNOLOHIYA and with the Southeast Asia Regional Institute for Community Education is linking concerned agronomists with KMP (the Peasant Movement of the Philippines) to find and adapt the old seeds to the new conditions. The task is not to freeze farmers in a traditional mode, but to allow farmer genius, natural evolution, and new scientific techniques to build upon each other for safer harvests in the future. By the mid-1980s, work was under way on forty-one varieties of rice. More must be collected. One community seed bank is in operation and others are planned.

Are the old varieties really a credible option? We spent a weekend bouncing along country roads with workers from the Agency for Community Education Services (ACES), a SIBAT affiliate organization. ACES was contracted by IRRI to study the impact of green revolution rice strains. Based on that study, people like Dinky Souman and George Villegas of ACES became convinced that if the cost of IRRI seed and corporate inputs remained as high as they had been, farmers would be better off with the old varieties. Detailed survey and test plot analysis bore out the conclusion, and it was endorsed by the national conference of farmers in 1985. As a result of ACES and SIBAT, scientists and peasants are working together to create their own agricultural revolution.

Promoting the conservation of traditional landraces and appreciating their value do not imply a wholesale rejection of modern varieties. We don't reject them. Few, if any, organizations that work with farmers reject them out of hand. That is not the issue.

Diversity, as we have often stated, is the raw material for evolution. And evolution is the prerequisite for survival. Some of this evolution takes place today in the fields of peasant farmers as they continue to practice the art of agriculture, selecting seeds and encouraging the adaptation of their seeds to their own ecological and social situations. And some of this evolution also takes place in the laboratories and test plots of scientists. Neither traditional nor modern varieties of seed are appropriate for all conditions. Each has its place. The world needs both. Thus, our discussion of the value of traditional varieties and the need to conserve them should not be misconstrued. The question has always been how to use both modern and traditional varieties properly, and how to conserve the traditional varieties in the face of the rapid advance of modern agriculture. Will the proper choices be made? Who will make them? Around the world people are facing these questions.

Long before anyone who writes books thought it was important, women in Zambia were talking with their elders and recording the use of herbs and wild vegetables, gathering the remaining grains and saving them for their children's children.[16] Long before there were FAO debates in Rome or battles over gene banks, villagers in Luzon were dividing up the responsibility for saving seeds: one responsible for cucumber; another for sweet potato; yet another for upland rice. School teachers in Changmai, Thailand, were organizing districts to grow old varieties on school property and Buddhist monks were working with village women to turn temples into seed sanctuaries. As the tempest of the green revolution has threatened to blow away diversity, farmers and gardeners from Ethiopia to Ecuador have sought to save seeds.

It is not an easy task. Poorly nourished people in the humid tropics might seem to be in the worst possible position to save seeds. Given the temperature, the rot, the insects and the hunger, it is hard to believe that a nongovernmental system could do much for genetic diversity. Villagers can hardly afford electricity and backup generators for cold storage units. In the villagers' favor, however, are their seeds and their determination to keep them.

Ethiopia's teff seed is highly durable under almost any conditions. Specialists at the Ethiopian Seed Corporation joke that teff could be left on a tabletop for twenty years and still germinate. In southeast Asia, farmers often save seed for three years—just in case—and the varieties survive.

Regardless of their keeping qualities, old varieties can still be hard to save. They take time and energy for people who have little to spare. They also take land. In the end, community seed storage, although comparatively cheap and efficient, is not guaranteed perfection. Some—perhaps a good deal—may still be lost.

A soft-spoken gardener in Decorah, Iowa, knows this well. Kent

Whealy launched his Seed Savers Exchange (SSE) in 1975 with no more support than the heritage of his grandfather-in-law's convictions—a handful of seeds entrusted to Whealy upon the man's death. Ten years later, Whealy has little money but a network of 630 North American farmers and gardeners in the (volunteer) business of growing out and sharing five thousand endangered vegetable varieties.[17]

The efforts of Whealy and the Seed Savers Exchange prove that even in the U.S. more than one genetic conservation system exists, for the SSE locates and conserves hundreds of varieties not found in any official government seed bank. In a 1985 study commissioned by the Office of Technology Assessment of the U.S. Congress, we found that the Seed Savers Exchange was conserving 1,799 "heirloom" varieties of beans. Of these, only 147 were to be found in government collections. The SSE had 544 varieties of tomatoes, of which the government could only locate 133. They had forty-six kinds of watermelons—the government was saving fewer than a quarter of them. And the Seed Savers Exchange was keeping five varieties of spinach, a crop for which seeds are difficult to save; the U.S. National Seed Storage Laboratory had none of the varieties in the SSE collection.[18]

The U.S. government collection has varieties not found in the SSE network, of course. But the point is that amateurs are conserving literally thousands of varieties of fruits and vegetables unknown to the government. And this indicates not only that more than one conservation system is operating (even in the U.S. with its relative poverty of diversity), but that more than one system is needed.

In the process of saving seeds, Whealy and his organization found it necessary to inventory all the public and private garden seed collections in North America. The first inventory surveyed the 1984 offerings of seed companies. Updating the original inventory in 1987, Whealy and his associates combed through 215 seed company catalogues listing over five thousand non-hybrid varieties.[19] SSE found an alarming rate of genetic erosion. Over half of the non-hybrid varieties were available from only one source (one of the 215 companies), meaning they were highly vulnerable to commercial extinction. More than two-thirds of the varieties were available from just two sources.

Every year, varieties are dropped from catalogues. In 1984, the year of the first inventory, 263 varieties were dropped from the lists.[20] Between 1984 and 1987, over nine hundred varieties were dropped from the catalogues. The Seed Savers Exchange set out to do something about it and began to buy up samples of any variety offered only by a single source. They have established the world's first "early warning system" for genetic erosion. Not satisfied with simply documenting the losses, SSE has a mechanism for obtaining and saving threatened varieties *before* they are lost. Working on a tight budget, this voluntary organization could teach a few lessons to the U.S. government's germplasm conservation system.

Sometimes preventive measures like those taken by the Seed Savers Exchange can be the last hope a species has. Not so long ago Easter Island's only tree species was thought to be extinct. The island's famous stone figures were probably rolled and moved using logs of this tree. Its wood may have been used in ancient times to make tablets on which the Polynesians' only written language was inscribed. But the last known tree died in 1962. Fortunately the director of the Göteborg Botanical Garden in Sweden came forth to report that the famous explorer and conservationist Thor Heyerdahl had recovered seeds from this tree on the *Kon-Tiki* voyage. Now the tree is being replanted on Easter Island.

EVERYBODY'S BUSINESS

When we were invited to talk to gene bank officials about the nongovernmental system of community seed "banks" in 1986 at a meeting called by IRRI, we were greeted with polite incredulity.[21] The whole idea seemed so disorganized. There was no clear definition of a community seed bank. To some it was a freezer or refrigerator; to others a farmer's seed bin; and to others yet, it was a piece of tropical forest or the apple trees in our backyard. In fact, the "bank" was more a concept than a physical entity. But how could governments control it? How could scientists work with it? Couldn't a thousand factors cause farmers suddenly to give up saving a crucial variety? How could we trust them? A few yards away was the world's most important collection of rice germplasm—80,000 accessions in the world's best gene bank serviced by the world's best scientists—in an earthquake zone.

Diversity. Choice. Options. Not one system but many. Not one basket. Not just national or international banks but community collections and biosphere reserves. Some reviewers of this book will surely find in these pages a condemnation of seed banks and those who work in them. It is difficult to promote an additional or complementary strategy and simultaneously avoid such misinterpretations. Just as both modern varieties and landraces have their legitimate place in various agricultural systems, both institutional and community based conservation strategies are also needed. There are reasons.

Governments have never been very good at eternity. In Nicaragua Humberto Tapia built a coffee gene bank, a living forest of choices. The dictator, Anastasio Somoza, used it to test diseases. Mario Gutíerrez collected maize diversity. But CIMMYT apparently threw it away. Botanists at Cornell University gathered apples. Cornell allowed the trees to be cut for firewood!

Jack Harlan—one of the true heroes of genetic diversity—once told us that, in the end, if diversity is to be saved, it may have to be saved by amateurs: people who love their seeds. And Harlan went on to add that throughout history, the amateurs had always been the ones to save diversity.

The people Daycha Siripat works with are such amateurs. A trained agronomist, Siripat heads the rice/fish project of the Appropriate Technology Association (ATA) of Thailand. Most of his time is spent in the villages in the northeast where he is working with CUSO and others to support local farmers in their efforts to bring fish back into the rice paddies. Before the green revolution, fish lived well in the irrigated paddies and served as a major source of protein and income for poor families. Modern rice varieties (even the Thai derivations of the IRRI originals) have brought with them demands for herbicides, insecticides, and fertilizers that have driven the fish away. With the collapse of rice markets, farmers in the northeast spontaneously began to bring back the old rice varieties and restock their paddies with fish.

Siripat and his organization have been helping by selecting the most useful fish species, subsidizing the original restocking, working with farmers to improve or restore old techniques—and seeking out the old seeds. The Appropriate Technology Association has gathered sixty varieties so far and they are actively searching for more. Early reports show that the fish fertilize the rice and cut down on the pest problems. Rice and fish yields are both soaring and family food and income have doubled among the most impoverished families. Siripat has searched through the national gene banks and is now looking to IRRI for more old varieties, but he is convinced that farmers will have to keep their own. The government search has been disappointing because hundreds of old rices have been "bulked" into single accessions. Community conservation—once again—is a necessity.[22]

In the final analysis, there is no one solution to the problem of genetic wipeout. Again, diversity applies. Parts of the solution are political; parts are technical; and parts are practical. No part is the answer by itself. Over the years of our work and research we have been forced to try to develop a constructive approach to this enormous problem. We've managed to consolidate our conclusions in five points.

FIVE LAWS OF GENETIC CONSERVATION

1. *Agricultural diversity can only be safeguarded through the use of diverse strategies.* No one strategy could hope to preserve and protect what it took so many human cultures, farming systems, and environments so long to produce. Different conservation systems can complement each other and provide insurance against the inadequacies or shortcomings of any one method.

2. *What agricultural diversity is saved depends on who is consulted. How much is saved depends on how many people are involved.* Farmers, gardeners, fishing people, medicine makers, religious leaders, carpenters—all have different interests that foreign scientists could never hope to appreciate fully. All segments of a community need to be involved to ensure that the total needs of the community are met. The more involvement, the greater the potential to conserve.

3. *Agricultural diversity will not be saved unless it is used.* The value of diversity is in its use. Only in use can diversity be appreciated enough to be saved. And only in use can it continue to evolve, thus retaining its value.

4. *Agricultural diversity cannot be saved without saving the farm community. Conversely, the farm community cannot be saved without saving diversity.* Diversity, like music or a dialect, is part of the community that produced it. It cannot exist for long without that community and the circumstances that gave rise to it. Saving farmers is a prerequisite of saving diversity. Conversely, communities must save their agricultural diversity in order to retain their own options for development and self-reliance. Someone else's seeds imply someone else's needs.

5. *The need for diversity is never-ending. Therefore, our efforts to preserve this diversity can never cease.* Because extinction is forever, conservation must be forever. No technology can relieve us of our responsibility to preserve agricultural diversity for ourselves and all future generations. Thus, we must continue to utilize diverse conservation strategies,

involve as many people in the process as possible, see that diversity is actively used and ensure the survival of the farm community—for as long as we want agricultural diversity to exist.

We present these five "laws" as a set of criteria for judging current and future efforts to conserve genetic diversity. These are the ingredients of a successful strategy. Anything less means that diversity either has not been saved or will not be conserved for long.

We realize that we have drawn a thoroughly bleak picture of the world's food security. More than once when discussing these issues with consumers and farmers, we have sensed a subtle shift toward the doorway as people instinctively edge in the direction of the nearest grocery or seed store. Are we about to lose rice? Will our next peanut butter sandwich be our last? Will next year or the year after be the Silent Spring? As we lose diversity, will we lose the capability of responding to the changes that the greenhouse effect will bring? Will we forfeit the sustainability of our agricultural system?

We honestly don't know. Intellect says the total loss of a major crop is technically possible and that current trends make such a loss almost inevitable. Instinct says that this will not happen: cataclysmic collapses are not as likely as continued gradual frittering away. We are much more likely to lose crops with a whimper than a bang.

Does that mean we will return to two wheatless days a week, like our grandparents had during World War I? Will fickle epidemics ravage our fields causing sudden famines and shortages? Predictions are risky. Crop losses may become more frequent and more severe but the worst disasters almost certainly will arise in the South. Some disasters will go unreported. In other cases, the world will only be treated to a body count and the cause will not be recorded.

Why in the South? In part because the North may have the needed genes in gene banks, and the scientific infrastructure to use them. In part because we in the North—or some of us—will buy our way out of hunger, and have something on the table. It may cost far too much. Growing it may sacrifice the environment. It may not be what we call food. It may afford the writers of recipes their greatest challenge since the hamburger. But there will be something to eat.

With imagination and commitment, a different future could be created—a future with more options, more choices, more life, more joy. It is a future we can already begin to discern.

THE GIFT OF COMMITMENT

To anyone who works to preserve genetic diversity, a visit to the Vavilov Institute in Leningrad is a pilgrimage. We arrived at our hotel in the early hours of a cool July evening in 1985. Not stopping to unpack, we threw our bags into the room and headed back down to the lobby, where an old doorman who spoke only Russian finally understood our pronunciation of "Vavilov." He pointed across the square in the direction of a large three-storey building. The offices were closed for the day and everyone had gone home, but there were flowers and exotic plants in the windows and we found a simple plaque commemorating Vavilov. This was the Vavilov Institute.

We had come for meetings with the director and staff of the institute, but as we walked through the front door and were confronted with a bust of Vavilov at the top of a staircase, business travel gave way to awe. At the base of the bust, staff members place flowers daily. The history of the place is very much alive. You cannot help but feel it.

The institute was preparing for the 1987 centenary of Vavilov's birth, when it would celebrate not only that event but its own history. It was from this institute that the first large-scale plant collections were launched. Vavilov and his associates searched the world for its crop diversity and brought back seeds to be stored in this building. Here Vavilov developed his theory of the centers of origin of agricultural crops, and the law of homologous series. And here is where he finished his career, a victim of Lysenkoism and ultimately a martyr for genetics.

Today his office looks much as it did the last time he left it in 1940: simple and practical. A big desk, overstuffed chairs, and some bookcases crammed with reports, old scientific instruments, and photographs from his many collecting expeditions. We found cytogeneticist Dr. Nina Tchuvashina leafing through some notebooks, preparing a history for the 1987 festivities. Our conversation, slowed by the necessity of going through an interpreter, turned to the Siege of Leningrad. The heroics of the people of Leningrad, surrounded and bombarded steadily by the Nazis for nine hundred days, are well-known. Less well-known is the story of the institute during the siege.

As the war began, institute scientists started to make duplicates of the 180,000-accession collection. Especially vulnerable was the potato collection, kept not as seed but as potatoes. Sub-zero temperatures would freeze them in the winter. Rats were a year-round threat. In the spring of 1942 the potatoes began to germinate, forcing scientists to plant them in the only place they had—in fields along the front. A valuable collection of blight-resistant potatoes made by Vavilov in Chile needed short days to mature and so was shielded from the sun by crude "cabins" constructed by the staff. By August, with the invaders near and the city in flames, institute staff began digging up the potatoes as shells hit the fields. Amazingly, they retrieved all the samples. During the winter as scientists were evacuated, they smuggled out potatoes sewn into pockets next to their bodies, so that the potatoes wouldn't freeze. All blight-resistant potatoes in the USSR today are descended from these potatoes.

The blockade of the city was forcing people to eat dogs, cats, rats, and even grass to stave off hunger. Over six hundred thousand people were to starve to death before the siege ended. Inside the institute, the rats had learned how to knock metal boxes full of seeds off the shelves in order to break them open. Guards were posted to protect the seeds from rats, and on the roof scientists took turns watching for fires caused by the shelling.

After the evacuation in 1942, thirty-one people were left at the institute. They were given a daily ration of 120 grams of bread (less than a quarter of a loaf of American bread). Fourteen died of starvation in December.

Dr. Tchuvashina brought out a scrapbook of photographs of these people and we sat around a table in Vavilov's outer office to look through it. There was Dr. Dmytry S. Ivanov, the rice specialist, who died at his desk surrounded by bags of rice. Dr. Rubtzov, a fruit breeder. As she leafed through the photographs, Dr. Tchuvashina paused over one; she said something to the interpreter who smiled and gave a short reply, before she turned to the next page. We asked the interpreter what she had said. She had asked him if he knew Dr. Geynts, the institute's librarian, now in his sixties. Yes, he did. The photograph was of Geynts' father—one of the men who had died of starvation in this building. We went on. Dr. Kreyer, a specialist in medicinal plants. Professor Molyboga, the meteorologist . . .

Why? Why would these people starve to death surrounded by so much food? Dr. Tchuvashina looked at us as if we must already know the answer—they were students of Vavilov. But what did they think they were doing saving all these seeds? What did they say to themselves as they slowly and collectively starved in this big old building? Dr. Tchuvashina reminded us that these scientists knew the value of genetic resources. Vavilov had taught them that. From where they were it looked as if humanity was destroying itself. Someday it would need these seeds. "When all the world is in the flames of war, we will keep this collection for the future of all people." This was what they were telling each other, she said.

We had no reply. We could only nod and thank her. As we walked across the square towards the hotel, our thoughts were on the sacrifices Vavilov and these scientists had made. Over four decades have passed since then. Plenty of time for us to learn to appreciate what these people gave their lives to save.

We have seen that appreciation in practice in Nicaragua. We have seen it at work in Ethiopia, where the government is preserving traditional varieties in anticipation of future droughts. We have talked with Thai farmers who insist on planting their native varieties of rice. We have even witnessed it in the rise of small companies and nonprofit organizations in North America, dedicated to promoting heirloom vegetable and fruit varieties. And we have watched this appreciation at work in the halls of the UN Food and Agriculture Organization as Third World ambassadors pursue agreements over germplasm exchange and the creation of an international gene fund.

Today we are not called upon to give our lives as the brave scientists at the Vavilov Institute did. We are not even required to be scientists or ambassadors, for remember it was the "amateurs" who domesticated our food crops and helped create diversity. Instead we are called upon to help preserve the diversity handed down to us. Whether we be scientists or politicians, farmers or factory workers, gardeners or teachers, we each have a special role to play in passing this gift on to the next generation. The manner in which we meet this challenge will largely determine how—or whether—future generations will live on this planet. "One thing is certain," writes Bentley Glass. "We cannot turn the clock back. We cannot regain the Garden of Eden or recapture our lost innocence. From now on we are responsible for the welfare of all living things, and what we do will mold or shatter our own heart's desire."[23]

PART THREE

Reference Material

NOTES

INTRODUCTION

1. Some reports state 1962 (see Doyle in *Altered Harvest,* for example) but IRRI's Tom Hargrove, who knows the researchers, claims that the first report was in 1961—from a conversation with Hargrove in August, 1986.

2. Doyle, Jack, *Altered Harvest* (New York: Viking, 1985), pp. 1–8, gives an excellent review of the impact and history of southern corn leaf blight.

3. National Academy of Sciences, *Genetic Vulnerability of Major Crops* (Washington, D.C., 1972), p. 13. Committee on Genetic Value of Major Crops of the National Research Council.

4. Private telephone conversation with Georgina Vitonova on August 12, 1986. Vitonova's loss estimate amounts to an average between figures from (see below) Hope Shand (25 percent) and James Trager (50 percent).

5. Trager, James, *Amber Waves of Grain* (New York: Arthur Fields, 1973), p. 13. Trager claims half the winter wheat was lost but Hope Shand of RAFI puts the figure closer to 25 percent. Shand gives the tonnage estimate.

6. From a personal conversation with a Canadian Wheat Board official on August 12, 1986.

7. "The World Grain Situation in 1972–1973," in *Foreign Agriculture,* USDA, October 23, 1972, p. 7.

8. Ibid., p. 9.

9. Demarco, Susan, and Susan Sechler, *The Fields Have Turned Brown,* The Agribusiness Accountability Project, 1975, p. 1. Washington, D.C.

10. Fischbeck, G., "The Usefulness of Gene Banks—Perspectives for the Breeding of Plants," in *UPOV Newsletter,* no. 25 (February 1981), pp. 15–16, from a symposium held October 15, 1980 by UPOV.

11. Personal communication from Dr. Robert Morrison, summer 1983, then a wheat breeder with Agriculture Canada at Lethbridge, Alberta.

12. Based on data provided by Dana G. Dalrymple, USAID, in September 1986, in documents and correspondence to Hope Shand of RAFI.

13. Schneider, Keith, "Researchers see gain in efforts to design crops," *New York Times,* March 16, 1986, p. 26. Schneider describes the start of biotechnology as "14 years ago" with the San Francisco experiments.

CHAPTER 1: ORIGINS OF AGRICULTURE

1. Holmberg, Uno, "Finno-Ugric, Siberian," in *The Mythology of All Races,* Canon John Arnott MacCulloch, ed., vol. 4 (Boston: Marshall Jones Co., 1927), p. 421.

2. Langdon, Stephen Herbert, "Semitic," in MacCulloch, *Mythology of All Races,* vol. 5, p. 103.

3. Harlan, Jack R., *Crops and Man* (Madison, WI: American Society of Agronomy, Inc. and Crop Science Society of America, Inc., 1975), p. 39.

4. Berndt, Ronald M., and Catherine H. Berndt, *Man, Land and Myth in North Australia* (East Lansing: Michigan State University Press, 1990).

5. Ibid., p. 7.

6. Lee, R.B., and I. DeVore, "Problems in the study of hunters and gatherers," in *Man the Hunter,* R.B. Lee and I. DeVore, eds. (Chicago: Aldine, 1968), pp. 3–12.

7. Mumford, Lewis, *The Myth of the Machine: Technics and Human Development* (New York: Harcourt, Brace, Jovanovich, 1966), p. 124.

8. Ibid., p. 79.

9. Dahl, Hartvig, *Word Frequencies of Spoken American English* (Detroit: Verbatim Books, Gale Research Co.), p. vii.

10. Harlan, Jack R., "A Wild Wheat Harvest in Turkey," *Archaeology,* vol. 20 (1967), p. 198.

11. Harlan, Jack R., and Daniel Zohary, "Distribution of Wild Wheats and Barley," *Science,* vol. 153 (September 2, 1966), p. 1079.

12. Ho, Ping-Ti, "The Loess and the Origin of Chinese Agriculture," *The American Historical Review,* vol. 75, no. 1 (October 1969), p. 34.

13. Lee, R.B., "What hunters do for a living, or how to make out on scarce resources," in Lee and DeVore, *Man the Hunter,* pp. 33–37.

14. Yanovsky, Elias, *Food Plants of the North American Indians,* USDA Miscellaneous Publication no. 237 (1936), p. 2.

15. Felger, Richard, and Gary Nabhan, "Deceptive Barrenness," *Ceres,* vol. 9, no. 2 (March–April 1976), p. 34.

16. Nabhan, Gary, personal correspondence, 1982.

17. Cohen, Mark N., "Population Pressures and the Origins of Agriculture: An Archaeological Example from the Coast of Peru," in *Origins of Agriculture,* Charles A. Reed, ed. (The Hague: Mouton Publishers, 1977), p. 140 (hereafter

cited as *Origins*). *Origins* is a major reference work on this subject. We relied upon it heavily in writing chapter 1.

18. Harlan, *Crops and Man,* p. 30.

19. Tannahill, Reay, *Food in History* (New York: Stein and Day Publishers, 1973), p. 20.

20. Hunn, Eugene S., "On the Relative Contribution of Men and Women to Subsistence Among Hunter-Gatherers of the Columbia Plateau: A Comparison with *Ethnographic Atlas* Summaries," *Journal of Ethnobiology,* vol. 1, no. 1 (May 1981), p. 132.

21. Harris, David, "Alternative Pathways Toward Agriculture," in Reed, *Origins,* p. 206.

22. Anderson, Edgar, *Plants, Man and Life* (Berkeley: University of California Press, 1967), p. 127.

23. Harlan, *Crops and Man,* p. 31.

24. Lee, in Lee and DeVore, *Man the Hunter,* p. 33.

25. Cohen, "Population Pressures and Origins of Agriculture," p. 138.

26. Harris, "Alternative Pathways," p. 213.

27. Reed, Charles A., "Introduction," in Reed, *Origins,* p. 44.

28. Flannery, Kent V., "Origins and Ecological Effects of Early Domestication in Iran and the Near East," in *Prehistoric Agriculture,* Stuart Struever, ed. (Garden City: Natural History Press, 1971), p. 60.

29. Harlan, Jack R., J.M.J. de Wet, and Ann Stemler, "Plant Domestication and Indigenous African Agriculture," in *Origins of African Plant Domestication,* Harlan, de Wet, and Stemler, eds. (The Hague: Mouton Publishers, 1976), p. 18.

30. Sauer, J.D., "Grain Amaranths," in *Evolution of Crop Plants,* N.W. Simmons, ed. (London: Longman Group Ltd., 1976), p. 6.

31. Harris, "Alternative Pathways," p. 229.

32. Mellaart, James, *Catal Huyuk: A Neolithic Town in Anatolia* (New York: McGraw-Hill, 1967), p. 82.

33. Harris, "Alternative Pathways," p. 213.

34. Ibid., p. 198.

35. Harlan, *Crops and Man,* p. 48.

36. Ibid., p. 17.

37. Flannery, "Domestication in Iran and the Near East," p. 95.

38. Shepard, Paul, *The Tender Carnivore and the Sacred Game* (New York: Scribner, 1973), p. 23.

39. Tannahill, *Food in History,* p. 46.

40. Bell, Barbara, "The Dark Ages in Ancient History. I. The first dark age in Egypt," *American Journal of Archaeology,* vol. 75 (1971), pp. 8–9.

41. Ibid., p. 12.

42. Harlan, J.R., J.M.J. de Wet, and E. Glen Price, "Comparative Evolution of Cereals," *Evolution,* vol. 27 (June 1973), p. 313.

43. Harlan, De Wet, and Stemler, *African Plant Domestication,* p. 7.

44. Lappé, F.M., *Diet for a Small Planet* (New York: Ballantine Books, 1971).

45. Harlan, De Wet, and Stemler, *African Plant Domestication,* p. 13.

46. Evans, L.T., "Crops and the World Food Supply," in *Crop Physiology,* L.T. Evans, ed. (Cambridge: Cambridge University Press, 1975), p. 2.

47. Frankel, O.H., "Variation Under Domestication," *Australian Journal of Science,* vol. 22, no. 1 (July 1959), p. 32.

48. Mehra, K.L., "Plant Genetic Resources: Their Nature and Priorities for Collection in South Asia," *Plant Exploration and Collection,* K.L. Mehra, R.K. Arora, and S.R. Wadhi, eds. (New Delhi: National Bureau of Plant Genetic Resources, 1981), p. 6.

49. Harlan, Jack R., "Evolution of Cultivated Plants," in *Genetic Resources in Plants,* O.H. Frankel and E. Bennett, eds. (London: International Biological Programme, 1970), p. 22 (hereafter cited as *Genetic Resources*). This book was the first thorough treatment of the subject. The authors used it extensively and consider it a true classic.

50. Fowler, Melvin, "The Origin of Plant Cultivation in the Central Mississippi Valley," in *Prehistoric Agriculture,* p. 122.

51. Harlan, De Wet, and Stemler, *African Plant Domestication,* p. 10.

52. Li, Hui-Lin, "The Vegetables of Ancient China," *Economic Botany,* vol. 23 (1969), p. 259.

53. Hedrick, U.P., ed., *Sturtevant's Edible Plants of the World* (New York: Dover Publications, Inc., 1972).

54. Coon, Nelson, *The Dictionary of Useful Plants* (Emmaus, PA: Rodale Press/Book Division, 1974).

55. Getahun, Amare, "The Role of Wild Plants in the Native Diet of Ethiopia," *Agro-Ecosystems,* vol. 1 (1974), pp. 47–48.

CHAPTER 2: DEVELOPMENT OF DIVERSITY

1. Rhoades, Robert, "The Incredible Potato," *National Geographic,* vol. 161, no. 5 (May 1982), p. 676.

2. Harlan, J.R., "Genetic Resources in Sorghum," *International Symposium of Sorghum in the 70s,* Ganga Prasada Rao and Leland R. House, eds. (New Delhi: Oxford and IBH Pub. Co., 1972), p. 1.

3. Harlan, *Crops and Man,* p. 164.

4. Harlan, De Wet, and Price, "Comparative Evolution of Cereals," p. 321.

5. De Wet, J.M.J., "Evolutionary dynamics of sorghum domestication," in *Crop Resources,* David S. Seigler, ed. (New York: Academic Press, 1977), p. 189.

6. Bosemark, N.O., "Genetic Poverty of the Sugarbeet in Europe," in proceedings of the July 3–7, 1978 conference *Broadening the Genetic Base of Crops,* A.C.

Zeven and A.M. van Harten, eds. (Centre for Agricultural Publishing and Documentation, Wageningen, Netherlands, 1979), p. 30 (hereafter cited as *Broadening the Base* proceedings).

7. Vavilov, N.I., "The Origin, Variation, Immunity, and Breeding of Cultivated Plants," *Chronica Botanica,* vol. 13, no. 1/6 (1949–50), p. 63.

8. Van Slambrouck, P., "Age-old cotton from Peru," *Christian Science Monitor* (March 3, 1981).

9. Haughton, Claire Shaver, *Green Immigrants* (New York: Harcourt Brace Jovanovich, 1979), pp. 433–434.

10. Van Rheenen, H.A., "Diversity of Food Beans in Kenya," *Economic Botany,* vol. 33, no. 4 (October–December, 1979), pp. 448–449.

11. Crisp, Peter, and George Forster, "Banking seeds for the future," *The Garden,* vol. 105, no. 10 (October, 1980), p. 410.

12. Smith, C.E., Jr., "Recent Evidence in Support of the Tropical Origin of New World Crops," in Seigler, *Crop Resources,* p. 93.

13. Bennett, Erna, personal correspondence, December 26, 1982.

14. Harris, David, "The environmental impact of traditional and modern agricultural systems," in *Conservation and Agriculture,* J.G. Hawkes, ed. (Montclair: Allanheld, Osmun and Co., 1978), pp. 67–68.

15. Harlan, J.R., and J.M.J. de Wet, "Some Thoughts About Weeds," *Economic Botany,* vol. 19, no. 1 (January–March, 1965), p. 20.

16. Anderson, Edgar, *Introgressive Hybridization* (New York: John Wiley and Sons, Inc., 1949), pp. 76–77.

17. De Wet, J.M.J., "II. Evolutionary dynamics of cereal domestication," (Part II of a Symposium on the Biochemical Systematics, Genetics and Origin of Cultivated Plants), *Bulletin of the Torrey Botanical Club,* vol. 102, no. 6 (November–December, 1975), p. 311.

18. Oldfield, Margery L., *The Value of Conserving Genetic Resources* (Washington, D.C.: U.S. Department of Interior, National Park Service, 1984), p. 75.

19. Burger, W.C., "Why Are There So Many Kinds of Flowering Plants?" *Bioscience,* vol. 31, no. 8 (September, 1981), p. 572.

20. Ehrlich, Paul and Anne Ehrlich, *Extinction* (New York: Random House, 1981), p. 94.

21. Burger, "Flowering Plants," p. 577.

22. Allard, R.W., "Population Structure and Sampling Methods," in Frankel and Bennett, *Genetic Resources,* p. 100.

23. Bradshaw, A.D., "Population structure, isolation and selection," in *Crop Genetic Resources for Today and Tomorrow,* O.H. Frankel and J.G. Hawkes, eds. (New York: Cambridge University Press, 1975), p. 41.

24. Bunting, A.H. and H. Kuckuck, "Ecological and Agronomic Studies Related to Plant Exploration," in Frankel and Bennett, *Genetic Resources,* p. 181.

25. Frankel, O.H. and Michael E. Soule, "The Genetic Diversity of Plants Used

by Man," *Conservation and Evolution* (New York: Cambridge University Press, 1981), p. 197.

26. Bennett, Erna, "Threats to Crop Plant Genetic Resources," in Hawkes, *Conservation and Agriculture,* p. 114.

27. Rhoades, "Incredible Potato," p. 694.

28. Ibid.

29. Smith, "Tropical Origin of New World Crops," pp. 87–88.

30. Bradshaw, "Population structure," p. 47.

31. Bennett, Erna, "Tactics of Plant Exploration," in Frankel and Bennett, *Genetic Resources,* p. 162.

32. De Wet, J.M.J., "Principles of evolution and cereal domestication," in *Broadening the Base* proceedings, p. 274.

33. Singh, H.B., "India," in *Survey of Crop Genetic Resources in their Centres of Diversity,* O.H. Frankel, ed. (Rome: Food and Agriculture Organization of the United Nations—International Biological Programme [FAO-IBP], 1973), p. 128 (hereafter cited as *Survey in Centres of Diversity*).

34. Rhoades, "Incredible Potato," p. 670.

35. Ragan, W.H., ed., *Nomenclature of the Pear: A Catalogue-Index of the Known Varieties Referred to in American Publications from 1804 to 1907,* USDA Bureau of Plant Industry Bulletin no. 126 (Washington, D.C.: Government Printing Office, 1926).

36. Haughton, *Green Immigrants,* p. 367.

37. Ibid., p. 143.

38. Brandolini, Aureliano, "Maize," in Frankel and Bennett, *Genetic Resources,* p. 294.

39. Vavilov, "Origin, Variation, Immunity," p. 58.

40. Heiser, Charles B., *Seed to Civilization—The Story of Man's Food* (San Francisco: W.H. Freeman & Co., 1973), p. 175.

41. Theophrastus, *Enquiry into Plants,* vol. 2, trans. Sir Arthur Hort (London: William Heinemann, Ltd., 1916), p. 167.

42. International Board of Plant Genetic Resources, *IBPGR Secretariat Consultation on the Genetic Resources of Cruciferous Crops,* 17–19 November 1980 (Rome, 1980), p. 8 (hereafter cited as *Resources of Cruciferous Crops*).

43. Parlevliet, J.E., J.G. Brewer, and W.G.M. Ottaro, "Collecting pyrethrum, *Chrysanthemum cinerariaefolium* Vis. in Yugoslavia for Kenya," in *Broadening the Base* proceedings, p. 91.

44. Yamashita, K., "Origin and Dispersion of Wheats with Special Reference to Peripheral Diversity," *Zeitschrift für Pflanzenzuchtung,* vol. 84, no. 2 (1980), p. 122.

45. Dobzhansky, T., "Soviet Biology and the Powers that Were," *Science* 164 (June 27, 1969), p. 1507.

46. Cohen, Barry Mendel, "Nikolai Ivanovich Vavilov—His Life and Work," (Ph.D. diss., University of Texas, Austin, 1980), p. 231.

47. Popovsky, Mark, *The Vavilov Affair* (Hamden: Archon Books, 1984), p. 186 ff.

48. Dobzhansky, T., "N.I. Vavilov, A Martyr of Genetics," *Journal of Heredity*, vol. 38 (August 1947), p. 230.

49. Zirkle, Conway, "L'Affair Lysenko: Spring 1956," *Journal of Heredity* 47 (March–April 1956), p. 47.

50. Cohen, Barry Mendel, "Vavilov—Life and Work," p. 220.

51. Dobzhansky, "Soviet Biology," p. 1508.

52. Harlan, J.R., "Our Vanishing Genetic Resources," *Science* (May, 1975), p. 621.

53. Brezhnev, D., "Mobilization, Conservation, and Utilization of Plant Resources at N.I. Vavilov All-Union Institute of Plant Industry, Leningrad," in *Genetic Resources*, p. 533.

54. Anderson, *Introgressive Hybridization*, p. 212.

55. Bennett, Erna, "Plant Introduction and Genetic Conservation," *Scottish Plant Breeding Station Record* (Pentlandfield, Roslin, Midlothian, 1965), p. 69.

56. Vavilov, N.I., *Studies on the Origin of Cultivated Plants* (Leningrad: Institut de Botanique Appliquée et d'Amélioration des Plates, 1926), p. 219.

57. Zohary, Daniel, "Centers of Diversity and Centers of Origin," in Frankel and Bennett, *Genetic Resources*, pp. 34–35.

58. Vavilov, "Origin, Variation, Immunity," pp. 20–43.

59. Wilkes, G., "Native Crops and Wild Food Plants," *Ecologist*, vol. 7, no. 8 (1977), p. 315.

60. Ho, Ping-Ti, "Loess and Chinese Agriculture," p. 26.

61. Rindos, David, "Symbiosis, Instability, and the Origins and Spread of Agriculture: A New Model," *Current Anthropology*, vol. 21, no. 6 (December, 1980), p. 757 ff.

62. Harlan, J.R., "Agricultural Origins: Centers and Noncenters," *Science*, vol. 174 (October 29, 1971), p. 473.

63. Frankel, "Variation Under Domestication," p. 32.

64. Rindos, "Symbiosis, Instability and Spread of Agriculture," p. 757 ff.

65. Yamashita, "Origin and Dispersion of Wheats," p. 129.

66. Hymowitz, T., and J.R. Harlan, "Introduction of Soybean to North America by Samuel Bowen in 1765," *Economic Botany*, vol. 37, no. 4 (October–December 1983), p. 373.

67. Bennett, "Plant Introduction and Genetic Conservation," p. 35.

68. Kupzow, A.J., "The Formation of Areas of Cultivated Plants," *Zeitschrift für Pflanzenzuchtung*, vol. 53, no. 1 (1965), p. 61.

69. Hutchinson, J.B., R.A. Silow, and S.G. Stephens, *The Evolution of Gossypoim* (Oxford: Oxford University Press, 1947), p. 88 ff.

70. Cooper, J. F., "Environmental Physiology," in Frankel and Bennett, *Genetic Resources*, p. 135.

71. Kupzow, "Areas of Cultivated Plants," p. 61.

72. Vavilov, *Origin of Cultivated Plants*, p. 204.

73. Purseglove, J.W., "The Origins and Migrations of Crops in Tropical Africa," in Harlan, De Wet, Stemler, *Origins of African Plant Domestication*, p. 299.

74. Cutler, Hugh C., "Medicine Men and the Preservation of a Relict Gene in Maize," *Journal of Heredity*, vol. 35, no. 10 (October, 1944), p. 290 ff.

75. Harlan, "Centers and Noncenters," p. 469.

76. Farney, Dennis, "Meet the men who risked their lives to find new plants," *Smithsonian* (June 1980), p. 134.

77. Jefferson, Thomas, *Thomas Jefferson: Garden Book*, ed. Edwin Morris Betts (Philadelphia: The American Philosophical Society, 1944), pp. 124, 131.

78. Purseglove, in Harlan, De Wet, and Stemler, *African Plant Domestication*, p. 380.

79. Myrdal, Gunnar, *Asian Drama*, vol. 1 (New York: Pantheon, 1968), pp. 442–443.

80. Rodney, Walter, *How Europe Underdeveloped Africa* (London: Bogle-L'Ouverture Publications, 1972), p. 181.

81. Harlan, J.R., "Genetics of Disaster," *Journal of Environmental Quality*, vol. 1, no. 3 (1972B), p. 212.

CHAPTER 3: VALUE OF DIVERSITY

1. Wilkes, G., "The World's Crop Plant Germplasm—An Endangered Resource," *The Bulletin of the Atomic Scientists* (February 1977), pp. 14–15.

2. Rhoades, Robert, "Incredible Potato," pp. 686–687.

3. Ibid., pp. 679, 682.

4. Kee, Robert, "Famine and the Fenians," *The Listener*, vol. 105, no. 2693 (January 1, 1981), p. 7.

5. Ibid.

6. Clark, Jack, review of *Paddy's Lament: Ireland 1846–1847*, "Prelude to Hatred," by Thomas M. Gallagher, *Food Monitor*, no. 31 (January/February 1983), p. 26.

7. Mitchel, John, *The Last Conquest of Ireland (Perhaps)*, (Dublin: Burns, Oats and Washburn), 1860, p. 150 ff.

8. Kee, "Famine and the Fenians," p. 7.

9. Ibid.

10. Ibid., p. 8.

11. Bennett, Erna, personal correspondence, September 3, 1988.

12. Mokyr, Joel, *Why Ireland Starved: A Quantitative and Analytical History of the Irish Economy, 1800–1850* (Boston: George Allen & Unwin, 1983), p. 42.

13. Stakman, E.C., and J.J. Christensen, "The Problem of Breeding Resistant Varieties," in *Plant Pathology,* J.G. Horsfall and A.E. Dimond, eds., vol. 3 (New York: Academic Press, 1960), p. 572.

14. Harris, David R., "The Environmental Impact of Agricultural Systems," in Hawkes, *Conservation and Agriculture,* p. 68.

15. Harlan, "Genetics of Disaster," p. 213.

16. Stakman and Christensen, "Breeding Resistant Varieties," p. 574.

17. Ibid., p. 578.

18. Browning, J.A., "Corn, Wheat, Rice, Man: Endangered Species," *Journal of Environmental Quality,* vol. 1, no. 3 (July–September 1972), p. 209.

19. United States Environmental Protection Agency, *Research Summary: Integrated Pest Management,* EPA-600/8-80-044 (Washington, D.C.: U.S. Government Printing Office, September 1980), p. 1.

20. May, Robert M., and Andrew P. Dobson, "Population Dynamics and the Rate of Evolution of Pesticide Resistance," in *Pesticide Resistance: Strategies and Tactics for Management,* National Research Council (Washington, D.C.: National Academy Press, 1986), p. 171.

21. Pyle, R.M., "Conservation of Lepidoptera in the United States," *Biological Conservation,* vol. 9 (1976), p. 71.

22. Brody, Jane E., "Farmers Turn to Pest Control in Place of Eradication," *Ag World* (September 1976), p. 7.

23. Carson, Rachel, *Silent Spring* (New York: Fawcett World Library, 1962), p. 239.

24. Forgash, Andrew J., "History, Evolution, and Consequences of Insecticide Resistance," *Pesticide Biochemistry and Physiology,* vol. 22 (1984), p. 184.

25. Food and Agriculture Organization Plant Production and Protection Paper, *Pest Resistance to Pesticides and Crop Loss, Assessment-2* (Rome, 1979), p. 16.

26. Chang, T.T., S.H. Ou, M.D. Pathak, K.C. Ling, and H.E. Kauffman, "The Search for Disease and Insect Resistance in Rice Germplasm," in Frankel and Hawkes, *Resources for Today and Tomorrow,* pp. 192–193.

27. Hills, Lawrence, Unpublished transcript of the Seeds Conference, held in Rome, 1981, by the International Coalition for Development Action (headquartered in Barcelona, Spain).

28. Harlan, J.R., "Genetic Resources of Some Major Field Crops in Africa," in Frankel, *Survey in Centres of Diversity,* p. 57.

29. Van der Plank, J.E., *Disease Resistance in Plants* (New York: Academic Press, 1968), p. 3.

30. Bennett manuscript, pp. 93–94.

31. Goring, C.A., "Chemical Technology: Witches' Brew or Mulligan Stew,"

Down to Earth trade magazine, Dow Chemical Company, vol. 35, no. 1 (Fall 1978), p. 1.

32. Hartley, William, "Climate and Crop Distribution," in Frankel and Bennett, *Genetic Resources,* pp. 143–144.

33. Bradshaw, David, "Plant Introductions Improve American Agriculture," *Agri-Search,* South Carolina Agricultural Experiment Station, Clemson University, vol. 1, no. 4 (Fall, 1980), p. 5.

34. Harlan, J.R., "Genetic Resources in Wild Relatives of Crops," *Crop Science* (May–June, 1976), p. 330 (hereafter cited as "Wild Relatives").

35. Harlan, J.R., Unpublished remarks at FAO/UNEP/IBPGR Technical Conference on Crop Genetic Resources in Rome (April 8, 1981), from author's notes.

36. Prescott-Allen, Robert and Christine, "Protected Areas and the Conservation of Genetic Resources," (Victoria, B.C.: PA DATA, 1979), pp. 2–3.

37. Ibid., p. 4.

38. Ibid.

39. Ross, H., "Wild Species and Primitive Cultivars as Ancestors of Potato Varieties," in *Broadening the Base* proceedings, p. 242.

40. Harlan, J.R., Unpublished remarks, FAO/UNEP/IBPGR conference.

41. Creech, John L., "Tactics of Exploration and Collection," in Frankel and Bennett, *Genetic Resources,* pp. 223–224.

42. Prescott-Allen, "Protected Areas," p. 8.

43. Ibid., pp. 6–7.

44. Rick, Charles, "Potential Improvement of Tomatoes by Controlled Introgression of Genes from Wild Species," in *Broadening the Base* proceedings, p. 170.

45. Chang, T.T., "The Case for Large Collections," in *The Use of Plant Genetic Resources,* A.H.D. Brown, O.H. Frankel, D.R. Marshall, and J.T. Williams, eds. (Cambridge: Cambridge University Press, 1989), p. 126.

46. Cooper, J.P., "Environmental Physiology," in Frankel and Bennett, *Genetic Resources,* p. 140.

47. Qualset, C.Q., "Sampling Germplasm in a Center of Diversity: An Example of Disease Resistance in Ethiopian Barley," in Frankel and Hawkes, *Resources for Today and Tomorrow,* p. 82.

48. Prescott-Allen, "Protected Areas," p. 9.

49. Ibid., p. 8.

50. Ibid., p. 6.

51. Ibid., p. 7.

52. Harlan, "Wild Relatives," pp. 330–331.

53. Lange, W., and G. Jochemsen, "Use of Wild Emmer (*Triticum dicoccoides,* AABB) in the Breeding of Common Wheat (*T. aestivum,* AABBDD)" in *Broadening the Base* proceedings, p. 225.

54. Harlan, "Wild Relatives," p. 331.

55. Ibid.

56. Prescott-Allen, Robert and Christine, *The First Resource: Wild Species in the North American Economy* (Boston: Yale University Press, 1988).
57. Culpeper, Roy, "The Debt Matrix," North-South Institute, Ottawa, April, 1988, p. 8.
58. Harlan, "Wild Relatives," p. 330.

CHAPTER 4: GENETIC EROSION

1. Stakman, E.C., R. Bradfield, and P.C. Mangelsdorf, *Campaigns Against Hunger* (Cambridge: Belknap Press, 1967), p. x.
2. Cleaver, Jr., H., *The Origins of the Green Revolution,* Ph.D. dissertation (Ann Arbor: Xerox University Microfilms, 1975), p. 244.
3. King, John, "Rice Politics," *Foreign Affairs* (April 1953), p. 453.
4. Cleaver, *Green Revolution,* p. 327.
5. Ibid., p. 309.
6. Ibid., p. 300.
7. Lappé, Francis Moore, Joseph Collins, with Cary Fowler, *Food First* (New York: Ballantine Books, 1978, revised edition), pp. 129 ff.
8. Griffin, Keith, and A.R. Khan, eds., *Poverty and Landlessness in Rural Asia.* A study by the World Employment Programme, manuscript (Geneva: ILO, 1976).
9. Griffin, Keith, "The Green Revolution: An Economic Analysis." (Geneva: United Nations Research Institute for Social Development, Report No. 72.6, 1972).
10. "Masagana 99," Ministry of Agriculture, The Philippines, Agriculture Extension, vol. 3, no. 8 (April 15, 1981), pp. 2–3.
11. Farvar, M. Taghi, "The Relationship Between Ecological and Social Systems" (Abstract), presented at the Panel on Inter-dependence of Ecosystems Around the World, Earthcare (UN: New York, June 6, 1975), p. 9.
12. Kuckuck, Hermann, "Importance of the Utilization, Preservation and Further Development of the National Genetic Resources for the Cultivation of Plants" in *Plant Research and Development* (Institute for Scientific Co-operation, Germany, 1975), p. 111.
13. Ruttan, V.W., "New Rice Technology and Agricultural Development Policy" in *Economic Consequences of the New Rice Technology* (Los Banõs, Philippines: International Rice Research Institute, 1978), p. 373.
14. Kuckuck, "Utilization, Preservation, Development," p. 112.
15. Goering, Theodore J., and World Bank, Agricultural Research: Sector Policy Paper. Washington, D.C.: World Bank, 1981, pp. 21–22.
16. Morgan, Dan, *Merchants of Grain* (New York: Viking Press, 1979), p. 9.
17. Frankel, "Variation Under Domestication," p. 30.

18. Chang, T.T., "Crop Genetic Resources," in *Plant Breeding Perspectives,* by D.J. van der Have (Wageningen, Netherlands: Center for Agricultural Publishing and Documentation, 1979), p. 84.

19. Goering, "Agricultural Research," p. 21.

20. Harlan, H.V. and M.L. Martini, "Problems and Results in Barley Breeding," *Yearbook of Agriculture, 1936* (Washington, D.C.: U.S. Department of Agriculture, 1936), p. 317.

21. The following are sources used to prepare inventories of fruit and vegetable varieties, and unpublished reports prepared by the Rural Advancement Fund International.

Fogle, H.W., and H.F. Winters, *North American and European Fruit and Tree Nut Germplasm Resources Inventory,* Miscellaneous Publication No. 1406, Washington, D.C.: U.S. Department of Agriculture, 1981, 732 pp.

National Seed Storage Laboratory, complete inventory on microfiche, provided by NSSL, Ft. Collins, Colorado, September, 1983.

Ragan, W.H., *Nomenclature of the Apple: A Catalogue of the Known Varieties Referred to in American Publications From 1804 to 1904,* Bureau of Plant Industry, Bulletin No. 56, Washington, D.C.: Government Printing Office, 1926.

Tracy, W.W., Jr., *American Varieties of Garden Beans,* Bureau of Plant Industry, Bulletin No. 109, U.S. Department of Agriculture, Washington, D.C.: Government Printing Office, 1907.

Tracy, W.W., Jr., *American Varieties of Lettuce,* Bureau of Plant Industry, Bulletin No. 69, U.S. Department of Agriculture, Washington, D.C.: Government Printing Office, 1904.

Tracy, W.W., Jr., *List of American Varieties of Vegetables for the Years 1901 and 1902,* Bureau of Plant Industry, Bulletin No. 21, U.S. Department of Agriculture, Washington, D.C.: Government Printing Office, 1907.

Chiosso, Elaine, "Vegetable Variety Inventory: Varieties from USDA 1903 List of American Vegetables in Storage at the National Seed Storage Laboratory," Unpublished report prepared by the Rural Advancement Fund International, Pittsboro, NC, 1983.

Rural Advancement Fund International, "Listing of Possible Extinct Varieties of Apples," Unpublished report prepared by RAFI, Pittsboro, NC, 1982.

Rural Advancement Fund International, "Listing of Possible Extinct Varieties of Pears," Unpublished report prepared by RAFI, Pittsboro, NC, 1982.

22. Hedrick, U.P., *The Pears of New York,* Report of the New York Agricultural Experiment Station (Albany: J.B. Lyon Co., 1921), p. 123.

23. Von Bothmer, Roland, and Niels Jacobsen, "*Hordeum* in North America," Report to IBPGR, 1982, p. 5. Available at IBPGR.

24. Frankel and Hawkes, *Resources For Today and Tomorrow,* p. 103.

25. Bennett, Erna, "Wheats of the Mediterranean Basin," in Frankel, *Survey in Centres of Diversity,* p. 4.

26. Singh, H.B., "India," in Frankel, *Survey in Centres of Diversity,* p. 127.

27. Frankel, Otto H., "Genetic Resources Survey as a Basis for Exploration," in Frankel and Hawkes, *Resources for Today and Tomorrow,* p. 106.

28. Porceddu, E., "Wheat Collecting in the Mediterranean Region," in *Broadening the Base* proceedings, p. 79.

29. Kuckuck, Hermann, "Present Situation of Genetic Resources of Small Grains in Syria, Iraq, Iran, Afghanistan and West Pakistan," Proceedings of Eucarpia Conference in Izmir (June, 1972), p. 76.

30. Porceddu, E., personal correspondence, 1983.

31. Frankel and Soulé, *Conservation and Evolution,* pp. 214–215.

32. Bennett, "Wheats of the Mediterranean," p. 6.

33. Kjellqvist, E., "Turkey," in Frankel, *Survey in Centres of Diversity,* p. 10.

34. Kuckuck, Hermann, *Report on a Survey of Genetic Resources of Small Grains Carried out in Syria, Iraq, Iran, West Pakistan and Afghanistan from May 10 to July 31, 1971,* p. 15.

35. Bennett, Erna, "Afghanistan," in Frankel, *Survey in Centres of Diversity,* p. 22.

36. ASSINSEL, *Feeding the 5000 Million,* International Association of Plant Breeders (Nyon, Switzerland: undated, unpaginated).

37. Harlan, J.R., "Seed Crops," in Frankel and Hawkes, *Resources for Today and Tomorrow,* pp. 114–115.

38. Frankel and Soulé, *Conservation and Evolution,* p. 191.

39. Myers, Norman, *The Sinking Ark* (Oxford: Pergamon Press, 1979), p. 61.

40. Pearce, Andrew, *Social and Economic Implications of the Green Revolution* (Oxford: Clarendon Press, 1980), p. 27.

41. Jain, H.K., "Plant Breeders Rights and Genetic Resources," *Indian Journal of Genetics,* vol. 42 (1982), p. 122.

42. Harlan, J.R., "How Green Can a Revolution Be?" in Seigler, *Crop Resources,* pp. 108–109.

43. Frankel, Otto H., "Genetic Conservation in Perspective," in Frankel and Bennett, *Genetic Resources* (see ch. 1, note 49), pp. 474–475.

44. Singh, "India," p. 135.

45. Frankel and Hawkes, *Resources for Today and Tomorrow,* p. 106.

46. De Wet, J.M.J., "Increasing Cereal Yields: Evolution Under Domestication," in Seigler, *Crop Resources,* p. 115.

47. Webster, O.J., "Sorghum Vulnerability and Germplasm Resources," *Crop Science,* vol. 16, no. 4 (July/August, 1976), p. 554.

48. Harlan, "Major Field Crops in Africa" (see ch. 1, note 29), p. 50.

49. Ochoa, Carlos, personal correspondence, 1983.

50. Ochoa, Carlos, "Native Potatoes in Bolivia, Chile and Peru," in Frankel, *Survey in Centres of Diversity,* p. 117.

51. Ochoa, Carlos, “Potato Collecting Expeditions in Chile, Bolivia and Peru, and the Genetic Erosion of Indigenous Cultivars,” in Frankel and Hawkes, *Resources for Today and Tomorrow,* p. 169.

52. Ochoa, “Potato Collecting Expeditions,” pp. 167–168.

53. Grubben, G.J.H., *Tropical Vegetables and Their Genetic Resources* (International Board for Plant Genetic Resources, Rome, 1977), pp. 111–113.

54. Hanelt, Dr. P., personal correspondence (1983).

55. Williams, J.T., and B.V. Ford-Lloyd, “Beet in Turkey,” *Plant Genetic Resources Newsletter,* no. 31 (August, 1975), p. 6.

56. Frankel, *Survey in Centres of Diversity,* p. 28.

57. Frankel and Soulé, *Conservation and Evolution,* p. 194.

58. Frankel, *Survey in Centres of Diversity,* p. 30.

59. Zagaja, S.W., “Temperate Zone Tree Fruits,” in Frankel and Bennett, *Genetic Resources,* p. 331.

60. Ibid., p. 332.

61. Crisp, Peter, personal correspondence, 1983.

62. Ibid.

63. Crisp, Peter, and George Forster, “Banking Seeds for the Future” (see ch. 2, note 11), p. 411.

64. International Board for Plant Genetic Resources, *IBPGR Secretariat Consultation on the Genetic Resources of Cruciferous Crops,* 17–19 November, 1980 (Rome, 1980), pp. 9–10.

65. Innes, N.L., “Genetic Conservation and the Breeding of Field Vegetables for the United Kingdom,” *Outlook Agriculture,* vol. 8, no. 5 (1975), p. 304.

66. IBPGR, *Resources of Cruciferous Crops,* p. 8.

67. Ibid., pp. 3–4.

68. Innes, “Conservation and Breeding of Field Vegetables,” p. 304.

69. Van der Meer, Q.P., personal correspondence, 1983.

70. Palmer, Ingrid, *Science and Agricultural Production* (United Nations Research Institute for Social Development, Geneva, 1972), pp. 95–96.

71. Van der Have, *Plant Breeding Perspectives,* p. 22.

72. Harlan, “Major Field Crops in Africa,” p. 49.

73. Rhoades, “Incredible Potato,” p. 694.

74. Myers, *Sinking Ark,* p. 62.

75. Rick, Charles, personal correspondence, 1983.

76. Rick, Charles, “Conservation of Tomato Species Germplasm,” *California Agriculture,* vol. 31 (1977), pp. 32–33.

77. Grubben, *Tropical Vegetables,* p. 20.

78. Ehrlich and Ehrlich, *Extinction,* p. 142.

79. Soria, Jorge, “Recent Cocoa Collecting Expeditions,” in Frankel and Hawkes, *Resources for Today and Tomorrow,* p. 178.

80. Chalmers, W.S., personal correspondence, 1983.

81. Soria, "Cocoa Collecting Expeditions," p. 175.
82. Soria, Jorge, personal correspondence, 1983.
83. Prescott-Allen, Robert and Christine, "Protected Areas," p. 10.
84. International Board for Plant Genetic Resources *Annual Report 1982* (IBPGR, Rome, 1983).
85. Bennett, Erna, "Historical Perspectives in Genecology," *Scottish Plant Breeding Station Record* (1964), p. 95.
86. Knott, D.R., and Dvorak, J., "Alien Germplasm As A Source of Resistance To Disease," *Journal of the Annual Review of Phytopathology,* vol. 14 (1976), p. 211.
87. Watson, I.A., "The Utilization of Wild Species in the Breeding of Cultivated Crops Resistant to Plant Pathogens," in Frankel and Bennett, *Genetic Resources,* pp. 452–453.
88. Bennett, "Historical Perspectives," p. 95.
89. Browning, J.A., "Corn, Wheat, Rice, Man" (see ch. 3, note 19), p. 210.
90. National Academy of Sciences, *Genetic Vulnerability of Major Crops,* p. 287.
91. Ehrenfeld, David W., *Conserving Life on Earth* (Oxford: Oxford University Press, 1972), p. 54.
92. Harlan, J.R., "Genetics of Disaster," p. 213.
93. Bosemark, N.O., "Genetic Poverty of the Sugarbeet in Europe," in *Broadening the Base* proceedings, p. 33.
94. Bennett, Erna, "Plant Introduction and Genetic Conservation," p. 82.
95. Ibid.
96. Mohan, S. Tara, and H.K. Jain, "Studies on Varietal Adaption in Wheat," *Zeitschrift für Pflanzenzuchtung,* vol. 76, no. 4 (June, 1976), p. 285.
97. Farney, Dennis, "Meet the Men Who Risked Their Lives," p. 129.
98. Haughton, *Green Immigrants,* pp. 436–437.
99. Bennett, "Plant Introduction and Genetic Conservation," p. 82.
100. Van der Have, *Plant Breeding Perspectives,* p. 91.
101. Ibid., p. 90.
102. Frankel, Otto H., "Genetic Dangers in the Green Revolution," *World Agriculture,* vol. 19, no. 3 (July 1970), p. 11.
103. Frankel, "Variation Under Domestication," p. 29.
104. Heslop-Harrison, J., "The Plant Kingdom: An Exhaustible Resource?" *Transactions of the Botanical Society of Edinburgh,* vol. 42 (1973), pp. 8–9.
105. RAFI collected U.S. Department of Agriculture and Extension Service Bulletins from major apple growing states and compared lists of apple varieties. See, for example, "Tree Fruit Varieties for Michigan," Cooperative Extension Service, Michigan State University, E. Lansing, Extension Bulletin E-881, June, 1983, and Way, R.D., "Apple Varieties Grown in New York State," *New York's Food and Life Sciences Bulletin,* No. 78, Geneva: N.Y. Agricultural Experiment Station, April, 1979.

106. Official Journal of the European Communities, Information and Notices, "Common Catalogue of Varieties of Agricultural Plant Species," Luxembourg: Office for Official Publications of European Communities, published annually. This publication is commonly referred to as the Common Catalogue.

107. Bennett, Erna, personal conversation, April, 1981.

108. Final Impact Statement, USDA, 1979, p. 2.

109. "The Last Word," Phil Donahue interview program, taped in Chicago on ABC-TV, January, 1983.

110. Harlan, "Genetics of Disaster," pp. 213, 215.

CHAPTER 5: TROPICAL FORESTS

1. Ehrlich and Ehrlich, *Extinction,* pp. 39–40.

2. Prance, Ghillean T., "The Amazon: Earth's Most Dazzling Forest," *Garden,* vol. 6, no. 1 (January/February, 1982), p. 9.

3. White, Peter T., "Tropical Rain Forests: Nature's Dwindling Treasures," *National Geographic,* vol. 163, no. 1 (January, 1983), p. 8.

4. Prance, Ghillean T., "The Amazon Forest: A Natural Heritage To Be Preserved" in *Extinction Is Forever,* Ghillean T. Prance and Thomas S. Elias, eds. (New York: The New York Botanical Garden, 1977), p. 673.

5. Gentry, Alwyn H., "Endangered Plant Species and Habitats of Ecuador and Amazonian Peru" in Prance and Elias, *Extinction Is Forever,* p. 136.

6. Ibid.

7. Holdridge, L.R., "Ecological and Genetical Factors Affecting Exploration and Conservation in Central America" in *Tropical Trees: Variation, Breeding and Conservation,* J. Burley and B.T. Styles, eds. (London: Academic Press, 1976), p. 199.

8. Raven, Peter, "Tropical Rain Forests: A Global Responsibility," *Natural History* (February, 1981), p. 28.

9. Myers, *Sinking Ark,* p. 134.

10. Ibid., p. 23.

11. Ibid., p. 131.

12. "The Infinite Voyage: Life in the Balance," produced by WQED/Pittsburgh in association with the National Academy of Sciences, March 1989, p. 5. Writer and producer, Joe Seamans.

13. Raven, "Tropical Rain Forests: A Global Responsibility," p. 28.

14. Ehrlich and Ehrlich, *Extinction,* p. 231.

15. Hughes, Carol and David, "Teeming Life of a Rain Forest," *National Geographic,* vol. 163, no. 1 (January, 1983), p. 53.

16. Rindos, "Symbiosis, Instability, and Spread of Agriculture," p. 753.

17. Goodland, R.J.A., and H.S. Irwin, *Amazon Jungle: Green Hell to Red Desert?* (New York: American Elsevier Pub. Co., 1975), p. 78.

18. Janzen, Daniel H., and Paul S. Martin, "Neotropical Anachronisms: The Fruits the Gomphotheres Ate," *Science*, vol. 215 (January 1, 1982), p. 19ff.

19. Goodland and Irwin, *Green Hell to Red Desert*, p. 108.

20. Prance, "The Amazon Forest," pp. 189–190.

21. UNESCO, *Tropical Forest Ecosystems*, a state-of-knowledge report prepared jointly by UNESCO/UNEP/FAO (Paris, 1978), p. 192.

22. Goodland and Irwin, *Green Hell to Red Desert*, p. 102.

23. Kemp, R. H., and J. Burley, "Depletion and Conservation of Forest Genetic Resources" in Hawkes, *Conservation and Agriculture*, pp. 179–180.

24. Raven, Peter, "Comments: Worldwide Needs and Opportunities" in *Proceedings of U.S. Strategy Conference on Biological Diversity*, November 16–18, 1981 (Dept. of State Publication 9262), pp. 55–56.

25. Lanly, Jean-Paul, *Tropical Forest Resources* (Rome: FAO Forestry Paper 30, 1982), p. 100.

26. Myers, *Sinking Ark*, p. 134.

27. Galeano, Eduardo, *Open Veins of Latin America: Five Centuries of the Pillage of a Continent* (New York: Monthly Review Press, 1973), p. 71.

28. Westoby, Jack C., "Halting Tropical Deforestation: The Role of Technology." Draft of an Office of Technology Assessment document, August 1982, p. 14.

29. Imle, Ernest P., "Hevea Rubber: Past and Future" in Seigler, *Crop Resources*, p. 119.

30. Ibid.

31. Brockway, Lucile H., *Science and Colonial Expansion: The Role of the British Royal Botanic Gardens* (New York: Academic Press, 1979), p. 157.

32. Imle, "Hevea Rubber," pp. 120–121.

33. Subramaniam, S., and Ong Seng Huat, "Conservation of Gene Pool in Hevea," *Indian Journal of Genetics and Plant Breeding*, vol. 34A, 1974, p. 33.

34. Bunker, Stephen G., "Forces of Destruction in Amazonia," *Environment*, vol. 22, no. 7 (September, 1980), p. 19.

35. Galeano, Eduardo, *Open Veins of Latin America*, p. 116.

36. Westoby, Jack C., personal communication, 1984.

37. Goodland and Irwin, *Green Hell to Red Desert*, p. 1.

38. Hecht, Susanna B., "Environment, Development and Politics: Capital Accumulation and the Livestock Sector in Eastern Amazonia," *World Development*, vol. 13, no. 6, 1985, p. 670.

39. Bunker, "Forces of Destruction," p. 35.

40. Ibid., p. 36.

41. Ibid.

42. Hecht, "Environment, Development and Politics," p. 670.

43. Bunker, "Forces of Destruction," pp. 39, 40.
44. Ibid., p. 38.
45. Lappé, Frances M., and Joseph Collins, with Cary Fowler, *Food First: Beyond the Myth of Scarcity* (New York: Ballantine Books, 1977), p. 50.
46. Nations, James D., and Daniel I. Komer, *Rainforest, Cattle, and the Hamburger Society* (Austin, Texas: Center for Human Ecology, 1982), p. 25.
47. Myers, *Sinking Ark*, p. 171.
48. Ibid., p. 141.
49. Hecht, "Environment, Development and Politics," p. 679.
50. Nations and Komer, *Rainforest, Cattle, and Hamburger*, pp. 24–25.
51. Myers, N., "The Hamburger Connection: How Central America's Forests Become North America's Hamburgers," *Ambio*, vol. 10, no. 1, 1981, p. 5.
52. Nations and Komer, *Rainforest, Cattle, and Hamburger*, p. 12.
53. Ibid.
54. Ibid., p. 6.
55. Ibid., p. 10.
56. Ibid., p. 8.
57. Myers, *Sinking Ark*, p. 146.
58. Ehrlich and Ehrlich, *Extinction*, pp. 163–164.
59. Nations and Komer, *Rainforest, Cattle, and Hamburger*, pp. 9–10.
60. Ibid., p. 7.
61. Ibid., p. 15.
62. "The Threat to the Tropical Rain Forests," Sveriges Television, Lasse Berg, producer, 1979.
63. Ehrlich and Ehrlich, *Extinction*, pp. 157–158.
64. Sveriges Television, "Threat to Tropical Forests."
65. Myers, *Sinking Ark*, p. 159.
66. Ibid., pp. 160–161.
67. Ibid., p. 158.
68. Bingham, C.W., *Multinational Issues: Some Observations and an Example* (Prepared for CNR 198 by C.W. Bingham, May 1973, Berkeley, California), p. 4.
69. Bunker, "Forces of Destruction," p. 16.
70. Westing, Arthur H., *Ecological Consequences of the Second Indo-China War* (Stockholm: Almqvist & Wiksel International, 1976), p. 10.
71. Westing, Arthur H., and E.W. Pfeiffer, "The Cratering of Indochina," *Scientific American*, vol. 226, no. 5 (May, 1972), p. 21.
72. Ibid.
73. Ibid., p. 24.
74. Westing, *Ecological Consequences*, pp. 51–52.
75. Westing and Pfeiffer, "Cratering of Indo-China," pp. 25–27.
76. Westing, *Ecological Consequences*, p. 47.

77. Ehrenfeld, *Conserving Life on Earth,* pp. 58–59.
78. Perry, Thomas O., "Vietnam: Truths of Defoliation," letter in *Science* (May, 1968), p. 601.
79. Westing, A. H., and C.E. Westing, "Endangered Species and Habitats of Vietnam," *Environmental Conservation,* vol. 8, no. 1 (Spring, 1981), p. 59.
80. Westing, *Ecological Consequences,* p. 30.
81. Ehrenfeld, *Conserving Life on Earth,* pp. 56–57.
82. UNESCO, *Tropical Forest Ecosystems,* p. 48.
83. Frankel, *Survey in Centres of Diversity,* p. 69.
84. Creech, John L., "Tactics of Exploration and Collection" in Frankel and Bennett, *Genetic Resources,* p. 226.
85. Bennett, Erna, "Plant Introduction and Genetic Conservation," p. 82.
86. Frankel, *Survey in Centres of Diversity,* p. 69.
87. International Board for Plant Genetic Resources, Report from Working Group on Genetic Resources of *Coffea arabica* meeting in Rome, December 11–13, 1979 (report issued 1980), p. 2.
88. Harlan, "Genetics of Disaster," p. 213.
89. Stakman, E.C., and J.J. Christensen, "Breeding Resistant Varieties," p. 571.
90. International Board for Plant Genetic Resources, "Bananas and Plantains," Report from Working Group on the Genetic Resources of Bananas and Plantains meeting in Rome, July, 1977 (report issued 1978), p. 5.
91. Ibid., p. 1.
92. Food and Agriculture Organization, *Forest Genetic Resources,* information, No. 9 (FAO, Rome, 1979), p. 8.
93. Withner, Carl L. "Threatened and Endangered Species of Orchids" in Prance and Elias, *Extinction Is Forever,* p. 317.
94. Shand, Hope, "Vanilla and Biotechnology," *RAFI Communique,* January, 1987, occasional reports published by Rural Advancement Fund International.
95. ASSINSEL, *Feeding the 5000 Million* (see ch. 4, note 33).
96. FAO, *Forest Genetic Resources,* p. 7.
97. Ehrlich and Ehrlich, *Extinction,* p. 71.
98. Gentry, "Endangered Species of Ecuador and Peru," p. 143.
99. Food and Agriculture Organization/United Nations Environment Programme, *Report on the FAO/UNEP Expert Consultation on In Situ Conservation of Forest Genetic Resources,* meeting held in Rome, December 2–4 (FAO, Rome, 1980), p. 33.
100. Palmberg, Christel, "Genetic Resources of Arboreal Fuelwood Species for the Improvement of Rural Living," Paper, FAO/UNEP/IPPGR Technical Conference on Crop Genetic Resources (Rome, April, 1981), p. 5.
101. FAO/UNEP, *In Situ Conservation of Forest Genetic Resources,* p. 33.
102. UNESCO, *Tropical Forest Ecosystems,* p. 48.

103. Ibid., p. 55.
104. Ehrlich and Ehrlich, *Extinction,* p. 91.
105. Myers, *Sinking Ark,* pp. 258–259.
106. UNESCO, *Tropical Forest Ecosystems,* p. 52.
107. Ibid., p. 55.
108. Myers, *Sinking Ark,* p. 259.
109. Prescott-Allen, Robert and Christine, "Protected Areas," pp. 5–16.
110. UNESCO, *Tropical Forest Ecosystems,* pp. 437–438.
111. Myers, *Sinking Ark,* p. 127.
112. Anonymous, "Medicinal Plant Lost?" *SCRIP-WORLD Pharmaceutical News,* October 1, 1986, p. 22.
113. Quinlivan, T., "The Amazon Indians," *Ceres,* vol. 15, no. 2 (March/April, 1982), p. 19.
114. Ehrlich and Ehrlich, *Extinction,* pp. 239–240.
115. Quinlivan, "The Amazon Indians," p. 19.
116. Ehrlich and Ehrlich, *Extinction,* pp. 239–240.
117. Quinlivan, "The Amazon Indians," p. 20.
118. Ehrlich and Ehrlich, *Extinction,* p. 240.
119. Barrett, Suzanne, "Conservation in Amazonia," *Biological Conservation,* vol. 18 (1980), p. 220.
120. Bingham, *Multinational Issues,* p. 18.
121. Chaney, William, and Malek Basbous, "The Cedars of Lebanon," *Economic Botany,* vol. 32, no. 2 (April–June 1978), p. 120.
122. Li, Hui-Lin, *The Origin and Cultivation of Shade and Ornamental Trees* (Philadelphia: University of Pennsylvania Press: 1963), p. 199.
123. Giordano, G., "The Mediterranean Region" in *A World Geography of Forest Resources,* S. Haden-Guest, J.K. Wright, and E.M. Teclaff, eds. (New York: Ronald Press, 1956), pp. 338–339.
124. Myers, *Sinking Ark,* p. 222.
125. Brune, A. and G.H. Melchior, "Ecological and Genetical Factors Affecting Exploitation and Conservation of Forests in Brazil and Venezuela" in Burley and Styles, *Tropical Trees,* p. 208.
126. Barrett, "Conservation in Amazonia," p. 225.
127. Myers, *Sinking Ark,* p. 228.
128. Kemp and Burley, "Depletion and Conservation of Forest Genetic Resources," p. 180.
129. Frankel and Soulé, *Conservation and Evolution,* pp. 121–122.
130. Harris, Richard, Lynn Maguire, and Mark Shaffer, "Sample Sizes for Minimum Viable Population Estimation," *Conservation Biology,* vol. 1, no. 1 (May, 1987), p. 72ff.
131. Frankel and Soulé, *Conservation and Evolution,* p. 123.
132. Newmark, William, "A Land-Bridge Island Perspective on Mammalian

Extinctions in Western North American Parks," *Nature,* vol. 325 (January 29, 1987), p. 430.

133. Frankel and Soulé, *Conservation and Evolution,* p. 117.

CHAPTER 6: RISE OF THE GENETICS SUPPLY INDUSTRY

1. National Academy of Sciences, *Genetic Vulnerability of Major Crops,* p. 1.

2. L. William Teweles & Co. of Milwaukee, Wisconsin, are seed and plant biotechnology industry consultants. Information obtained from company brochure and advertising.

3. FAO, "World List of Seed Sources" (Rome: FAO/AGP: SIDP/82/5 November, 1982), pp. 121–122.

4. Kent, James W., "The Driving Force Behind the Restructuring of the Global Seeds Industry," *Seed World,* June, 1986, p. 25.

5. Ibid.

6. Anonymous, "ICI Likes Stauffer's Chemistry," *Businessweek,* June 22, 1987, p. 54.

7. From a confidential discussion with EEC officials in Brussels, October, 1981. See also James W. Kent in *Seed World,* June, 1986, p. 26, for a comparison with pharmaceuticals and petrochemicals.

8. FAO, "World List of Seed Sources," p. iii.

9. Organization for Economic Cooperation and Development, "OECD Schemes for the Varietal Certification of Seeds Moving in International Trade—List of Cultivars Eligible for Certification" in 1987 (Paris, September, 1988), Appendix VI, pp. 64–83.

10. FAO, "World List of Seed Sources."

11. Telephone communication with CGIAR Secretariat in Washington, D.C., October, 1987.

12. Private communication from Dr. Martin Kenney, Ohio State University, May, 1988.

13. Smith, Marvanna, Chronological Landmarks in American Agriculture, U.S. Department of Agriculture, May, 1979, p. 26.

14. In a 1978 letter from Carl Buchting to the president of the Canadian Seed Trade Association.

15. News release, Clayton & Dubilier, December 30, 1986, notes takeover by General Foods in 1971, ITT in 1979, and leveraged buyout via Clayton & Dubilier in 1986. Burpees and O.M. Scott sales were $200 million then.

16. Groosman, Ton, Anita Linnemann, and Holke Wierema, "Technology Development and Changing Seed Supply Systems," Research Report No. 27, Seminar Proceedings, June 22–23, 1988, pp. 17–18, Instituut Voor Ontwikkelings-Vraagstukken, Tilburg University.

17. Thomas, James H., Chief of Party, Consortium for International Develop-

ment, in a letter to Vic Althouse, Canadian Member of Parliament, January 13, 1981.

18. Williams, J.T., and B.V. Ford-Lloyd, "Beet in Turkey," IBPGR *Plant Genetic Resources Newsletter,* no. 31 (September, 1975), p. 3.

19. Crisp, Peter, and Brian Ford-Lloyd, "A Different Approach to Vegetable Germplasm Collection," IBPGR *Plant Genetic Resources Newsletter,* no. 48 (December, 1981), p. 11.

20. Frankel and Soulé, *Conservation and Evolution,* p. 272.

21. National Academy of Sciences, *Genetic Vulnerability of Major Crops,* p. 1.

22. Walton, Declan J., Director, IAA, in an FAO Inter-office Memorandum to L.E. Huguet, Director, FOR, February 28, 1980, "Subject: Cooperation with the Int'l. Union for the Protection of New Varieties of Plants."

23. *Plant Breeding Perspectives,* J. Sneep and A.J.T. Hendriksen, eds. (Wageningen: Centre for Agricultural Publishing and Documentation, 1979), p. 406.

24. Crittenden, Ann, "Gene Splicing and Agriculture," *New York Times,* May 5, 1981, p. D2.

25. Anonymous, *Business Week,* November 21, 1970, p. 30.

26. Palmer, Ingrid, *Science and Agricultural Production* (Geneva: UNRISD, 1972), p. 6 ff.

27. From discussions with Jerry van Kouverton of CUSO-Thailand, working with paddy rice farmers in Thailand, in August, 1986.

28. Comments made by the audience during a speech by Pat Mooney organized by the Finnish United Nations Association, March, 1982.

29. From personal discussions with Dr. Richaria, who had been director of agricultural research in India at the time of the green revolution, in Batu Malang, Indonesia, December, 1987.

30. From discussion with Dr. Melaku Worede of Plant Genetics Resource Center of Ethiopia in Addis Ababa, in June, 1987.

31. From communication with Dr. Martin Kenney, Ohio State University, May, 1988.

32. Letter from Dr. M.H. Hahmoud, Senior Agronomist, Ministry of Agriculture, Sudan, to Vic Althouse, Canadian Member of Parliament, May 5, 1981.

33. From a conversation with Dr. Yilma Kabede in Addis Ababa in July, 1985. Dr. Kabede met with Ciba-Geigy officials a few weeks earlier. The company, however, did leave samples with the Ethiopian gene bank.

34. Agricultural Institute of Canada, "Plant Breeders' Rights," 1979, p. 22.

35. The German advertisement was reproduced in Mooney, Pat Roy, "The Law of the Seed: Another Development in Plant Genetic Resources," *Development Dialogue,* 1983: 1–2, pp. 127–128.

36. de Ponti, O.M.B., comments appearing in unpublished paper. He also made these comments in a personal conversation with author, November, 1980, during debate at Wageningen.

37. Mirov, Kurt Rudolf, and Harry Maurer, *Webs of Power: International Cartels and the World Economy* (Boston: Houghton Mifflin, 1982), p. 121.

38. Kugel, Y., and G.W. Gruenburg, *International Payoffs* (Lexington: Lexington Press, 1977), p. 78.

39. Ibid.

40. Mirov, K.R., *Webs of Power,* p. 121.

41. Congress of the United States, Office of Technology Assessment, *Pest Management Strategies in Crop Protection* (Washington, D.C.: Government Printing Office, October, 1979), p. 72.

42. *Pest Management Strategies,* p. 72.

43. Silvey, Valerie, "The Contribution of New Varieties to Increasing Cereal Yield in England and Wales," National Institute of Agricultural Botany, United Kingdom, 1978, pp. 277–282.

44. The authors met with Dr. Keith Doodson in his office during a visit to Cambridge, U.K., in 1980.

45. Wormell, Peter, "Are High Yields Necessary?" *Agri-Trade News* (London, May, 1981), p. 28.

46. Economic Commission for Europe, "The Influence of Environmental Protection Measures on the Development of Pesticide Production and Consumption," United Nations Economic and Social Council (1982), p. 83.

47. Kloppenburg, Jack R., *First the Seed: The Political Economy of Plant Biotechnology, 1492–2000* (New York: Cambridge University Press, 1988), p. 13.

48. Quoted in Ryder, Edward J., "The Art and Science of Plant Breeding in the Modern World of Research Management," *HortScience,* vol. 19, no. 6 (December, 1984), p. 810.

49. Statement made by Byron Bealer, vice-president of Ciba-Geigy Seeds, to the Canada Grains Council in the spring of 1977, during a discussion on plant breeders' rights held in Winnipeg, Manitoba.

50. Office of Technology Assessment, *Genetic Technology—A New Frontier* (Westview Press, 1981), p. 204.

51. Kenney, Martin, *Biotechnology: The University-Industrial Complex* (New Haven: Yale University Press, 1986).

CHAPTER 7: ENTER BIOTECHNOLOGY

1. Anonymous, "Tobacco Plant Glows with Firefly Gene Implant," *Applied Genetics News,* vol. 7, no. 5 (December 1986), pp. 10–11.

2. Anonymous, "Splice Anti-Freeze Gene Into Plant," *Ag Biotechnology News* (July/August, 1987), p. 21.

3. Reid, William J., "Biotechnology and Breeding Team Up in Agriculture," *Bio/Technology,* vol. 5, no. 9 (September, 1987), p. 903.

4. Sharp, W.R., "Opportunities for Biotechnology in the Development of New Edible Vegetable Oil Products," *Journal of the American Oil Chemistry Society,* vol. 63, no. 5 (May, 1986), p. 598.

5. Anonymous, "Japan Roundup," *Bio/Technology,* vol. 6, no. 11 (November, 1988), p. 1276.

6. Ratafia, Manny, and Terry Purinton, "World Agricultural Markets," *Bio/Technology,* vol. 6, no. 3 (March, 1988), p. 281.

7. Jones, L.H., "Biotechnology in the Improvement of the Oil Palm," in P.K. Stumpf, J.B. Mudd, and W.D. Nes, eds., *The Metabolism, Structure and Function of Plant Lipids* (New York: Plenum Press, 1987), p. 677.

8. Larson, Russell E., "Cocoa Raw Product-Production and Problems," in P.S. Dimick, ed., *Proceedings of the Symposium Cacao Biotechnology* (University Park: Penn State University, 1986), p. 7.

9. See *Genetic Contributions to Yield Gains of Five Major Crop Plants,* W.R. Fehr, ed., Crop Science Society of America, Special Publication No. 7, 1984.

10. Levin, Simon A., "Safety Standards for the Environmental Release of Genetically Engineered Organisms," *Tree,* vol. 3, no. 4; *Tibtech,* vol. 6, no. 4 (April, 1988), p. S49.

11. Ellstrand, Norman C., "Pollen as a Vehicle for the Escape of Engineered Genes?" *Tree,* vol. 3, no. 4; *Tibtech,* vol. 6, no. 4 (April, 1988), p. S31.

12. Ehrenfeld, David, "Commentary: Implementing the Transition to a Sustainable Agriculture: An Opportunity for Ecology," *Bulletin of the Ecological Society of America* (1987), p. 7.

13. Ellstrand, "Pollen as a Vehicle," p. S31.

14. Proceedings of the National Academy of Sciences, U.S.A., vol. 82 (March, 1985), p. 1406.

15. Jones, Dr. L.H., quoted in "Biotechnology: A Young Industry with Potential," *Journal of American Oil Chemist's Society,* vol. 64, no. 9 (September, 1987), p. 1230.

16. Personal discussion in July, 1985, Addis Ababa, Ethiopia.

17. Peeters, John P., and Nick W. Galwey, "Germplasm Collections and Breeding Needs in Europe," *Economic Botany,* vol. 42, no. 4 (October–December, 1988), p. 503ff.

CHAPTER 8: GLOBAL CONSERVATION BEGINS

1. Based on notes taken from T.T. Chang's file on the Beltsville meeting and from conversations with Chang at IRRI in the Philippines, August, 1986.

2. The authors obtained their information for this discussion (pp. 86–88) from many sources, some confidential. Among those who can be identified are Erna Bennett, T.T. Chang, and Oscar Brauer. All information is from personal conversations.

3. Plucknett, D.L., N.J.H. Smith, J.T. Williams, and N. Murthi Anishetty, "Crop Germplasm Conservation and Developing Countries," *Science*, vol. 220 (April 8, 1983), pp. 163–167.

4. Authors' notes from verbal presentation by IBPGR chairman to the FAO Committee on Agriculture, March 1983.

5. Plucknett, "Crop Germplasm Conservation." IBPGR staff size has since increased with a number of contract positions.

6. Ibid., as well as information obtained from discussions with scientists from the countries involved, including conversations with germplasm workers in Addis Ababa at the PGRC/E First International Symposium, October 13–15, 1986.

7. From a conversation with Dr. Nigel Smith, consultant to CGIAR, held at the University of California/Los Angeles, May 1986.

8. From numerous interviews with Erna Bennett during the 1980s.

9. Lyman, Judith M., "Progress and Planning for Germplasm Conservation of Major Food Crops," IBPGR *Plant Genetic Resources Newsletter*, no. 60 (December 1984), pp. 3–21. Technical Advisory Committee Secretariat, CGIAR, "Report of the Second External Program and Management Review of the International Board for Plant Genetic Resources (IBPGR) for the Consultative Group on International Agriculture Research Technical Advisory Committee, 36th Meeting, Rome, 11–18 March, 1985," FAO, Rome, February 1985, AGR/TAC: IAR/85/1 (restricted).

10. Technical Advisory Committee Secretariat, CGIAR, "Report of the Second External Program and Management Review of the International Board for Plant Genetic Resources," p. 31 for TAC/CGIAR figures only.

11. Personal conversation with Erna Bennett, Rome, 1981. This is now widely recognized as a problem and has led to IBPGR proposals for ecogeographical surveys and multidisciplinary collection expeditions.

12. Crisp, Peter, and Brian Ford-Lloyd, "A Different Approach to Vegetable Germplasm Collection," IBPGR *Plant Genetic Resources Newsletter*, no. 48 (December, 1981), p. 11.

13. Williams, J.T., "Gene Banks and Global Repositories," in *UPOV Newsletter*, no. 25 (February, 1981), p. 6.

14. International Conference on Crop Genetic Resources, Proceedings (FAO/UNEP/IBPGR: Rome, 1981), p. 5.

15. *Nature*, December 2, 1982.

16. IBPGR Annual Report, 1983, p. vii.

17. From a telephone conversation with Dr. Roland Loiselle of the Ottawa Gene Bank on June 1, 1978.

18. Prasada Rao, K.E., and M.H. Mengesha, "Sorghum and Millets Germplasm Collection in Eastern Sudan," ICRISAT Progress Report, Genetic Resources, no. 16, p. 2.

19. Personal communication from Melak H. Mengesha, Leader, Genetic Re-

source Unit, ICRISAT, to Vic Althouse, Member of Parliament in Canada, dated September 27, 1982.

20. Ibid., p. 8.

21. Prescott-Allen, Robert and Christine, "*In Situ* Conservation of Crop Genetic Resources," Draft for Review, AGP:IBPGR/81/10 January 1981, p. 16, with reference to maize collections in Latin America.

22. Report, IBPGR Advisory Committee on *Phaseolus* Genetic Resources meeting at CIAT in Cali, Colombia, September 6–7, 1976 (AGP:IBPGR/76/24), p. 3.

23. Ibid., p. 16 for other bean examples.

24. Prescott-Allen, "*In Situ* Conservation," p. 16.

25. This appears to have been the case for maize material at CIMMYT in the seventies when the collection was allowed to deteriorate.

26. "Interim Report on Working Group Convened by the Board to Advise on Action Needed on the Genetic Resources of Forage Plants," May 12–13, 1979 (AGP:IBPGR/79/49), p. 4.

27. Bill Brown made the statement in Chicago on October 18, 1984 at a symposium sponsored by Earthscan, in which we also participated.

28. Report of the TAC Quinquennial Review of the International Board for Plant Genetic Resources, Rome, 1980, AGP/TAC:IAR/80/2, Rev. 1 (restricted), p. 79; and personal comments from team members.

29. Report of the National Agricultural Research and Extension Users Advisory Board, USDA (October 1980), p. 6.

30. "The Department of Agriculture Can Minimize the Risk of Potential Crop Failures," Report to the Congress by the Comptroller General, General Accounting Office, April 10, 1981 (CED-81-75), pp. iv–v.

31. Ibid., p. iii.

32. Crittenden, Ann, "US Seeks Seed Diversity as Crop Assurance," *New York Times,* September 21, 1981, p. A1.

33. Tao, K.L., Travel Report, April 12–May 3, 1984, pp. 1–2. This is an unnumbered internal IBPGR staff report.

34. Ibid., p. 3.

35. From the meeting minutes of the Expert Committee on Plant Genetic Resources (Canada), November 24, 1977, p. 8.

36. *Regina Leader-Post,* November 2, 1986.

37. Crittenden, A., "Seed Diversity and Crop Assurance."

38. Murata, Minoru, Eric Roos, and Takami Tsuchiya, "Chromosome Damage Induced by Artificial Seed Aging in Barley. I. Germinability and Frequency of Aberrant Anaphases at First Mitosis," *Canadian Journal of Genetics and Cytology,* vol. 23, no. 2, 1981, p. 268.

39. Ibid., p. 267.

40. Ibid., p. 277.

41. Roos, Eric, "Genetic Shifts in Mixed Bean Populations. II. Effects of Regeneration," *Crop Science,* vol. 24, no. 4 (July/August 1984), p. 715.

42. Roos, Eric, "Genetic Shifts in Bean Populations During Germplasm Preservation," Annual Report of the Bean Improvement Cooperative, no. 20, 1977, p. 47.

43. The original confidential draft was available in January 1985, but the polished version was not available until a year later. We will quote from the confidential study.

44. Ibid., pp. 7–8.

45. Ibid., p. 8.

46. Ibid., p. 1.

47. Chapman, C.G.D., "Wheat Genetic Resources: A Review and Proposals for Future Action," AGPG:IBPGR/85/10 (restricted), p. 1.

48. Lyman, Judith M., "Progress and Planning for Germplasm Conservation of Major Food Crops," in IBPGR *Plant Genetic Resources Newsletter,* no. 60, (December 1984), AGP/PGR/60, p. 6.

49. Ibid., p. 6.

50. Williams, J.T., "Gene Banks and Clonal Repositories," in *UPOV Newsletter,* no. 25 (February, 1981), pp. 6–7, reprinting text of his speech to a UPOV Symposium on The Use of Genetic Resources in the Plant Kingdom.

51. This is our own estimate based on a study of IBPGR collection reports.

52. IBPGR Ad Hoc Advisory Committee on Seed Storage, unpublished report on first meeting, September 10–11, 1981, AGP:IBPGR/81/73, p. 8.

53. Report of IBPGR Working Group on Engineering, Design and Cost Aspects of Long-term Seed Storage Facilities, IBPGR, Rome, AGPE:IBPGR/76/25 December 1976, pp. 6–7.

54. IBPGR Ninth Board Meeting, "IBPGR Ad Hoc Advisory Committee on Seed Storage" Prov. Agenda No. 9.1, report by E.H. Roberts, AGP:IBPGR/81/73, no page number.

55. Hanley, Charles J., "Arctic Seeds," *Associated Press,* September 24, 1987.

56. Done, Kevin, "Permafrost Politics," *Financial Times,* October 10, 1986.

CHAPTER 9: POLITICS OF GENETICS RESOURCE CONTROL

1. Buckley, James L., "Welcome and Introduction," in *Proceedings of the U.S. Strategy Conference on Biological Diversity,* Washington, D.C., November 16–18, 1981, p. 10.

2. Murray, Dr. James R., "Biological Diversity and Genetic Engineering," in *Proceedings of the Strategy Conference,* pp. 40–44, with some quotes from unpublished portions of the full speech.

3. Murray, "Biological Diversity and Genetic Engineering," pp. 40–44.

4. Garbini, Giovanni, *The Ancient World* (New York: McGraw-Hill, 1966), pp. 122–123 for confirmation. Visit was made by Pat Mooney and Marilyn McGregor in June 1976.

5. Smith, Nigel J.H., "Botanic Gardens and Germplasm Conservation," University of Hawaii, Harold L. Lyon Arboretum, Lecture no. 14, 1986, p. 8.

6. Ibid.

7. *Seedsmen's Digest,* Editorial, 1985.

8. Lomni, Hely, "Is the Patent System Applicable to Biotechnological Inventions?" *UPOV Newsletter,* no. 54 (May, 1988), pp. 42–43.

9. Consumer and Corporate Affairs Canada, "Working Paper on Patent Law Revision," 1976, p. 10, quotes Emperor Zeno in 480 A.D. opposing monopolies on anything—expressly garments and fish.

10. Lemmon, Kenneth, *Golden Age of Plant Hunters* (Cranbury: A.S. Barnes and Co., 1968), p. 15.

11. Ibid.

12. Brockway, Lucile H., *Science and Colonial Expansion: The Role of the British Royal Botanical Gardens* (New York: Academic Press, 1979), p. 39.

13. Haughton, *Green Immigrants,* pp. 76–77.

14. Brockway, *Science and Colonial Expansion,* p. 112.

15. Brockway, Lucile H., "Plant Science and Colonial Expansion—The Botanical Chess Game," in Jack R. Kloppenburg, Jr., ed., *Seeds and Sovereignty* (Durham: Duke University Press, 1988), pp. 49–64.

16. Wood, David, "Crop Germplasm: Common Heritage or Farmers' Heritage?" in Kloppenburg, *Seeds and Sovereignty,* pp. 274–288.

17. Sondahl, M.R., et al., "Coffee" in P.V. Ammirato, et al., eds., *Handbook of Plant Cell Culture, vol. 3: Crop Species* (MacMillan, 1984), p. 566.

18. Brockway, *Science and Colonial Expansion,* p. 51.

19. Ibid., p. 52.

20. Brockway, "The Botanical Chess Game," p. 55.

21. Brockway, *Science and Colonial Expansion,* pp. 86–87.

22. Smith, "Botanic Gardens and Germplasm Conservation," p. 12.

23. The authors have visited all of these locations. Additional information derived from Brockway, *Science and Colonial Expansion,* and Smith, "Botanic Gardens and Germplasm Conservation," University of Hawaii, Harold L. Lyon Arboretum, Lecture no. 14, February 6, 1985.

24. Haughton, *Green Immigrants,* p. 418—but beware; we find this history unduly romantic and often exaggerated.

25. Brockway, *Science and Colonial Expansion,* p. 2. Actually, she suggests the date of 1759. George III played at Kew as a child.

26. IARCs are found in the Philippines, India, Syria, Ethiopia, Kenya, Nigeria, Liberia, Peru, Colombia, and Mexico. MIRCENs are found in Argentina, Brazil,

Canada, Egypt, Guatemala, India, Kenya, Senegal, Sweden, U.K., and U.S.A.—among others.

27. Kirsop, Barbara, "Tissue Culture Collections—Their Service to Biotechnology," *Trends in Biotechnology,* vol. 1, no. 1, 1983; editorial, "Microbiological Research Centres," *Science,* July 25, 1986, p. 401.

28. Dalrymple, Dana, *Development and Spread of High-Yielding Wheat Varieties in Developing Countries* (Washington, D.C.: U.S. Agency for International Development, 1986), p. 96.

29. Ibid.

30. Goodman, Major M., "Exotic Maize Germplasm: Status, Prospects and Breeding," *Iowa State Journal of Research,* vol. 59, no. 4 (May 1985), p. 501.

31. Ibid., p. 504.

32. Dalrymple, *High-Yielding Rice Varieties,* p. 115. Figures derived from text.

33. Data on crop value and USAID grants was provided by Dana Dalrymple of USAID in September, 1986, in a private communication to Hope Shand of RAFI.

34. Dalrymple, Dana, *High-Yielding Rice Varieties,* pp. 115–116.

35. *Bio/Technology,* vol. 5, no. 5 (May, 1987), p. 426.

36. Dias, Clarence J., and Yas P. Ghai, "Plant Breeding and Plant Breeders Rights in the Third World: Perspectives & Policy Options," International Development Research Centre (IDRC) Draft Report (April 1983), p. 19.

37. Assistant Deputy Director-General of IRRI in a dinner conversation at IRRI, August, 1986, referring to hybrid rice work by OXY, Cargill and Ciba-Geigy in Asia.

38. From *Diversity,* vol. 1, no. 3 (November/December, 1982), p. 9.

39. IBPGR Annual Report, 1987. Data has been collected from table 1 (pp. 29–32) and table 2 (p. 35).

40. Data calculated from Donald L. Plucknett, Nigel J.H. Smith, J.T. Williams, and N. Murthi Anishetty, *Gene Banks and the World's Food* (Princeton, N.J.: Princeton University Press, 1987), pp. 110–141.

41. Prescott-Allen and Prescott-Allen, *The First Resource,* especially pp. 198–203.

42. Data for the World "Network" Crop column in table 5 is interpreted from Plucknett, et al., *Gene Banks and the World's Food,* pp. 110–141, or, if crop data was unavailable from this source, from the appropriate IBPGR "Directory of Germplasm Collections" wherein RAFI has tallied the crop collection held by each country.

43. From IBPGR Annual Report, 1987, pp. 29–32 and 35. In most cases, the U.S. shares "global" status with one other nation or more. "Regional" implies a mandate for any region of the world (e.g., "New World").

44. Wood, David, in a letter addressed to "Genetic Resources Units: IARCs," Annex to the letter, September 15, 1987, emphasis in original.

45. Minutes, Fifteenth Meeting of IBPGR Board of Trustees, Rome, February 24–26, 1988 (IBPGR/88/46), Restricted to Board Members, p. 5.

46. Figures are derived from IBPGR reports of expeditions financed and numbers of accessions obtained. The cost of collections varies dramatically depending upon the species and the remoteness of the locale. In discussion with informed scientists, the authors have concluded that these are average figures for the mid-1980s.

47. Plucknett, et al., "Crop Germplasm Conservation," confirms embargo, as do Ethiopian officials.

48. Information from FAO/Committee on Agriculture Report, 1983, "freely available within India only."

49. From conversations with Cenargen authorities in May, 1984, and other discussions with the Malaysian Rubber Research Institute in July, 1985.

50. FAO/COAG Report.

51. From IBPGR germplasm data book on industrial crops in which Firestone rubber collection is described as "restricted."

52. Many officials in Brazil, Malaysia, India, Indonesia, and Nicaragua have made this point. Daniel Querol, then in Nicaragua, claims that Unilever was charging heavily for basic germplasm.

53. Plucknett et al., "Crop Germplasm Conservation," as well as other communications between Williams and the IIRB (trade association).

54. From FAO/UNDP/IBPGR Survey of European gene banks, 1982.

55. Information from the Canadian gene bank in Ottawa.

56. Stated in a letter from Dr. Thomas Curren, Research Officer, Science and Technology Division, Research Branch, Library of Parliament, to Vic Althouse, Canadian Member of Parliament, dated May 29, 1984.

57. Letter from T.W. Edminster, Administrator, Agricultural Research Service, U.S. Department of Agriculture, to Mr. Richard H. Demuth, Chairman, IBPGR, dated January 19, 1977.

58. Letter 8H16.1—this number was assigned by the U.S. Government in fulfilling its legal obligation to provide the document to the Foundation for Economic Trends and other participants in the legal action.

59. Letter dated December 7, 1983 to Dr. H.M. Munger, Cornell University professor.

60. From a telex from the U.S. Embassy in Rome to the U.S. State Department dated only "September, 1984" and designated by the U.S. Government as "14U530." This telex was among many surrendered by the U.S. Government to Jeremy Rifkin of the Washington, D.C.-based Foundation for Economic Trends as part of a legal action.

61. From a telex sent by the U.S. Embassy in The Hague, Netherlands, to the U.S. State Department dated only "October, 1984" and designated by the U.S.

Government as "14U489" as part of the same legal action noted above. This telex contains a letter from Dr. J. Hardon to the Embassy.

62. From a note by Paul J. Fitzgerald, Agricultural Science Advisor, U.S. Department of Agriculture, dated July 25, 1986, and addressed to Dr. Henry Shands. As part of the same legal action mentioned above, the U.S. Government has designated the document as "175345.2."

63. Memorandum from Dr. Henry L. Shands dated Februai, 12, 1987, and designated as document "178345.1" in the above-mentioned legal action.

64. Telephone conversation with Leon Weber of the Rodale Amaranth Project on September 13, 1988.

65. Kloppenburg, Jack, Jr., and Daniel Kleinman, "The Common Bowl: Plant Genetic Interdependence in the World Economy," AAAS, Philadelphia, May 28, 1986, p. 23 (unpublished). Figures are adapted by RAFI.

CHAPTER 10: RESPONSIBILITY AND COMMITMENT

1. Fowler, Cary, Eva Lachkovics, Pat Mooney, and Hope Shand, "The Laws of Life: Another Development and the New Biotechnologies," *Development Dialogue*, 1988, 1–2, pp. 194–209. For a more complete discussion of biological warfare and a more extensive listing of sources, see "On Mars and Microbes," written by the staff of the Rural Advancement Fund International.

2. Harris, Robert, and Jeremy Paxman, *A Higher Form of Killing* (New York: Hill and Wang, 1982), p. 103.

3. McDermott, Jeanne, *The Killing Winds: The Menace of Biological Warfare* (New York: Arbor House, 1987), p. 128ff.

4. Douglass, Joseph, and Neil Livingstone, *America the Vulnerable: The Threat of Chemical/Biological Warfare* (Lexington: Lexington Books, 1987), p. 33.

5. Personal discussion in July, 1985, in Addis Ababa.

6. From a personal conversation with Dr. Yilma Kebede in Addis Ababa in July, 1985, in the offices of Dr. Melaku Worede.

7. Krauss, Adolf, "Collection to Rescue Germplasm in the Drought Affected Areas of Ethiopia," Plant Genetic Resources Center/Ethiopia—International Livestock Center for Africa *Germplasm Newsletter*, no. 4 (December 1983), p. 6.

8. Personal discussion with Ian Rossiter, a World Bank official at the Ethiopian Seed Corporation, July, 1985. Rossiter strongly supports the regional concept but could not get backing in Washington.

9. FAO Seed Review, 1979–80, Seed Improvement and Development Programme, FAO, Rome, 1981, p. 14.

10. Ibid., p. 15. While six improved cultivars are listed, the study identifies four as very important.

11. Ibid. There are nine new cultivars but one is listed as "very important."

12. Personal conversation in Addis Ababa in July, 1985.

13. From a conversation with the chairman of the Bilateral Food Aid Committee (of donor countries) in Addis Ababa in February, 1985. The authors visited Ethiopia twice in 1985 and again in 1986, 1987, 1988, 1989, and 1990, and have been first-hand witnesses to the conservation effort.

14. From a discussion with Daniel Querol in Santiago, Chile, in October, 1988.

15. From a telephone conversation with Brian Tomlinson, Central American desk officer for the Canadian University Service Overseas, Ottawa, in 1985. The bombing incident was not reported in the world press because Canadian aid rules prevent financial support for projects arising from wars. Tomlinson asked the Rural Advancement Fund International (RAFI) to assist with obtaining alternative vegetable seeds.

16. From a conversation with Jean Christie, Executive Director, Inter Pares (a Canadian nonprofit development association based in Ottawa), who discussed the results of a women's workshop with the authors in 1986.

17. Whealy, Kent, and Arllys Adelmann, eds., *Seed Saver's Exchange: The First Ten Years, 1975–1985* (Decorah: Seed Saver Publications, 1986), p. 3.

18. Fowler, Cary, "Report on Grass Roots Genetic Conservation Efforts." Commissioned paper for U.S. Congress Office of Technology Assessment, 1985.

19. Whealy, Kent, ed., *The Garden Seed Inventory* (Decorah: Seed Saver Publications, 1985). The Garden Seed Inventory is a 448-page inventory of seed catalogues, listing all non-hybrid vegetable and garden seeds still available in the U.S. and Canada.

20. Ibid., pp. 7–8.

21. Pat Mooney was invited to IRRI in August–September, 1986, to give a series of lectures to a training workshop for gene bank directors (sponsored by IRRI) and to address the staff of IRRI. Mooney spoke extemporaneously and, to our knowledge, was not recorded. Reference to his visit was made in the 1986 IRRI annual report.

22. From personal discussion with Daycha Siripat and Gerry van Koeverden on September 3, 1986, in Bangkok, and from a paper by Kenneth T. MacKay et al., "Rice-Fish Culture in North East Thailand," IFOAM, August, 1986.

23. Glass, Bentley, *Science and Ethical Values* (Chapel Hill: University of North Carolina Press, 1965).

ACKNOWLEDGMENTS

Books are social products. They build upon the work of others. This is certainly true with *The Threatened Gene*. Without the rich history created by others, we would never have got the ideas or information with which to write this book. For inspiration in the early days of this work, we are deeply indebted to Erna Bennett, Lynn Randels, and Jack Harlan.

A collection of friends too numerous to name saw us through the long process of writing, which began a decade ago. Each has a special place in our hearts. We wish particularly to thank Natalie Hubbard and Marilyn McGregor.

Many people assisted us with research. From the very beginning until the sweet end, we had the skilled and sensitive assistance of Hope Shand. Her research, editing, and personal support were more important to the process than we could ever acknowledge here. We hope she realizes this.

Susie Crate, Laurie Heise, Elaine Chiosso, and Alice Ammerman did important and innovative research for us. We also thank our trusted colleague Eva Lachkovics and our fellow RAFI staff members Harald Wosihnoj, Silvio Martins, and Jose R. Manna de Deus.

The workhorses, however, were the people who provided secretarial and clerical support. We will never forget the cheerful and skilled help and the dedication of Beverly Cross and Tracy Strowd. They kept us organized and reasonably sane, though at times they must have had their doubts. We also thank Tema Okun, Betty Bynum, Diane Childs, Kathy Zaumseil, and Sean Peers.

Many people read the manuscript, offered their comments and gave editorial assistance. Each offered a unique perspective. Gary Nabhan worked with the entire manuscript in several drafts and offered feedback that was both tough and sensitive. Jack Harlan, one of our real heroes in this work, made comments on the first few chapters. Erna Bennett turned her critical eye to the manuscript and in the process challenged us immensely. What a different book this would be were it not for the help of these friends. The late Jack Westoby provided much of the theoretical basis for our chapter on tropical forests.

Our friends Bland Simpson, Anne Fitzgerald, Joy Bannerman, and Adele Negro helped edit the manuscript and improved it immeasurably.

Cary's parents, Judge and Mrs. Morgan Fowler, as well as Alan and Laura Haney and Joel and Melba Goldsby provided refuges to which we escaped to work during critical periods.

For their various contributions we wish to thank Kent Whealy, Rene Salazar, Tim Brodhead, Robert Morrison, Nelson Coyle, Helmut Kuhn, Jean Christie, Jeremy Rifkin, Anwar Fazal, Martin Abraham, Gay Wilentz, Ross Mountain, Lawrence Hills, Camila Montecinos, Kristin McKendall, Didi Soetomo, Erna Whitolar, Andrew Mushita, Vandana Shiva, Daniel Querol, Francisco Martinez, Rudiger Stegemann, Melaku Worede, friends at FAO, Gina Burkhardt, Jack Doyle, Henk Hobbelink, Dennis Lavalle, Sara Arnold, Harris Gleckman, Allen Tallos, and Drs. Julian Rosenman and Pac Heinsley and the staff of the University of North Carolina Memorial Hospital.

This book was supported by the Rural Advancement Fund and the International Coalition for Development Action. During the last few years both authors have been employed by the Rural Advancement Fund International. People in these organizations exhibited sometimes unreasonable amounts of trust and faith in the authors. Without the support of Kathryn Waller and the board and staff of the Rural Advancement Fund, this book would not have been published in this century. They know how true this is!

Our work was financially (and personally) supported by the CS Fund. In particular we wish to thank Maryanne Mott, Herman Warsh and Marty Teitel. We remain indebted to them for their continuing assistance and for the partnership which has evolved over the years.

We received important support from the Ruth Mott Foundation, Inter Pares, Phil Stern and the Stern Fund, the Jessie Smith Noyes Foundation, the Right Livelihood Foundation, the Canadian Council for International Cooperation, the Saskatchewan Council for International Cooperation, and Agnes Lindley. The Dag Hammarskjold Foundation helped in ways too numerous to list. Thank you Sven Hamrell, Olle Nordberg, and all of your colleagues at Geijersgarden overlooking Castle Hill. We are encouraged and humbled by the support of these fine people.

Thanks are also due to our literary agent Fran Collin of the Rodell-Collin Literary Agency. She too had great faith and commitment, for she understood early in the game that the financial rewards would be few.

Finally, we hope that we have properly acknowledged contributions others have made. None of those listed above or cited in the text should be held responsible for any mistakes. We have received help from many. But the conclusions we have reached and any mistakes we may have made are our own.

INDEX

ABOUT THE AUTHORS

Cary Fowler and Pat Mooney have traveled extensively since 1975, working with government officials, scientists, and grass-roots organizations on six continents on issues concerning the conservation, control, and use of genetic resources. In 1985 they received The Right Livelihood Award (the alternative Nobel Prize) in the Swedish Parliament for their "untiring and inspiring work" to alert the world to the dangers of losing our plant genetic diversity.

Both authors are on the staff of the Rural Advancement Fund International, a nonprofit organization working for a just and sustainable agriculture. Their work, in particular, focuses on the use, control, and conservation of genetic resources. Fowler works in Pittsboro, North Carolina, and Mooney in Brandon, Manitoba. Mooney is author of *Seeds of the Earth,* while Fowler is coauthor with Frances Moore Lappé and Joseph Collins of *Food First: Beyond the Myth of Scarcity.*

Countries visited by the authors while writing this book include all of Western Europe and Scandinavia (except Portugal), the USSR, Ethiopia, Egypt, Sudan, Kenya, Tanzania, Zimbabwe, Zambia, Uganda, Senegal, Pakistan, India, Bangladesh, Sri Lanka, Burma, Thailand, Malaysia, Indonesia, Singapore, Philippines, China, South Korea, Taiwan, Hong Kong, Japan, Mexico, Nicaragua, Chile, Brazil, Uruguay, the United Kingdom, Ireland, Australia, the United States, and Canada.